Between a Rock
and a Hot Place

lit·er·a·cy part·ners

Between a Rock and a Hot Place

Why Fifty Is *Not* the New Thirty

Tracey Jackson

HARPER

An Imprint of HarperCollins*Publishers*
www.harpercollins.com

The author is not a medical doctor and the advice given in this book is for informational purposes only. Consult your personal health advisor before following any of the advice herein.

HarperCollins books may be purchased for educational, business, or sales promotional use. For information, please write: Special Markets Department, HarperCollins Publishers, 10 East 53rd Street, New York, NY 10022.

FIRST EDITION

Designed by Eric Butler

Library of Congress Cataloging-in-Publication Data
Jackson, Tracey.
Between a rock and a hot place : why fifty is *not* the new thirty / Tracey Jackson.—1st ed.
p. cm.
ISBN 978-0-06-166927-9
1. Middle-aged women—Psychology. 2. Aging. 3. Jackson, Tracey. I. Title.
HQ1059.4.J32 2011
305.244'2—dc22
2010028725

11 12 13 14 15   OV/RRD   10 9 8 7 6 5 4 3

For Glenn, Taylor, and Lucy
For Taylor, Lucy, and Glenn
For Lucy, Glenn, and Taylor
So you know I love you all the same . . .

Contents

Between a Rock
and a Hot Place

Introduction

Fifty is the new thirty" has become the mantra of delusional baby boomers unwilling to relinquish their youth. Endless magazine articles, photos, and even T-shirts declaring this absurd theory have been appearing ever since the first of the millions of boomers started celebrating their fiftieth birthday with a colonoscopy and a membership card from AARP.

Considering that by the year 2025 there will be more than thirty million Americans over the age of sixty-five, boomers are at the beginning of the not so long, not remotely slow march into old age. Many boomers are already there. Those of us who are fifty are the youngsters in the boomer group. "Old age," according to most charts, starts at sixty. You can retire at fifty-five, you can move into a retirement community (also known as a senior citizens' community or the more euphemistic mature persons' community) once you turn fifty, and many people can start collecting their pensions in their early fifties.

So despite the chorus of boomers yelling, "We're actually

thirty!" the universe's compass seems to be pointing us directly toward old age, and it will be only about five years before many of us get there.

The image most of us have of being over fifty—or old, if you will—is our grandparents. Regardless of how much we loved them, most of us find the notion of turning into them intolerable.

Based on this fact alone, you can begin to see how the delusional concept arose. Thousands of boomers (iPlugged up the wazoo) started looking at photos of Grandma and Grandpa. "Wow," a boomer says while scanning the photos on his iPhone or iPad, trying to find something awesome to upload to his Facebook or Tumblr page. A shot breezes by of Grandpa, decked out in flood pants and a bowling shirt (not retro; his league), all topped off with a Mr. Rogers sweater and those soft, squishy, good-for-your-arches shoes. Grandma is standing over a pie; she's wearing a housedress and no makeup, and she's got more wrinkles than a shar-pei. The boomer, wearing some expensive jeans that keep ending up in his eighteen-year-old's drawer (they bought the same style), a vintage tee, and Converses, can't believe his Lasiked eyes. He does the math (easy, since the calculator is right there—I mean, did Steve Jobs know our needs or what?) and breathes, "Shit, they were fifty when that was taken. I'm fifty. If they looked like that at fifty and I look like this, that totally makes me thirty!" And thus another notch was added to the urban myth belt.

"Second adulthood," "third act," "the golden years," "final lap." Pick your euphemism—what we are slowly turning is old, and though the years may move at the same pace they always have, people always say the last lap feels the shortest. And we are not happy about it. We are so unhappy about it we are collectively pretending it's not happening. But I'm not sure denial is the right approach.

The other day I was at my gym (where I go six days a week, as I am committed to keeping whatever part of my "thirty"-year-old body I possibly can) when a woman who works at the gym—tall, thin, and hard-bodied in a way only someone whose daily work clothes are made of spandex can be—started talking about her upcoming birthday. I couldn't help asking, "What birthday is it?"

Attempting to cover her distress, she mumbled meekly, "Fifty."

Whew, we were the same age. She had a much better ass, but I had far fewer crow's feet.

She then immediately added, "It doesn't matter, because after all, fifty is the new thirty.

"How?" I retorted.

She stared at me blankly.

"Really, how is it the new thirty? Will you travel in a time machine and go back to 1987? Are you going to leave your husband and start dating thirty-two-year-old guys? Perhaps you can reintroduce yourself to the job market and share an apartment with three other girls. Will you stop getting mammograms and go back on the pill?"

Her endorphin-induced tranquility vanished in her panic; she started to sweat.

"I guess you'll stop getting hot flashes too."

She turned a dusty pink.

"It's just not true!" I yelled. "It's a lie, probably started in a bar by a fifty-year-old guy who was trying to pick up a thirty-year-old girl. And then some poor fifty-year-old divorcee sitting with a bunch of other fifty-year-old divorcees who were all being ignored because the fifty-year-old guys were hitting on the thirty-year-old girls, well, they heard it—out of context, mind you, but it gave them hope, and when you're fifty and sitting invisibly in a bar, you'll grab on to any affirmation you possibly can."

I understood where she was coming from because for a long time I believed it too. It was the comfy, cozy blanket I would wrap around my terror when I thought about getting older. And I'm sorry, but really and truly, no one wants to get older. The truth is that "older" (which, if you are lucky, moves eventually into "old") means one thing: closer to the end. It's incontrovertible. Most of us are afraid of death, and nobody wants to fall apart. So I don't buy that getting older is something people look forward to, embracing it with open arms and wallowing in all its glories. It's a lovely thought—acceptance of the golden years as a time full of the wisdom of experience, the understanding of what is important, and (oh God, I hear this all the time) "Now I really know who I am and it feels so good." And my other favorite, "I'm so happy I have nothing to prove." There are ways of making the best of the situation at hand. I'm a big believer in that. But I would bet my house that if you gave people the chance to freeze-frame themselves at, say, forty-three, they

4

would. I know I would. It's a great age; you've got enough of that wisdom, probably a good hunk of what you will have at sixty-five. You know yourself well enough not to make the big mistakes, and your body, if you work it, can still be pretty kick-ass. Your joints are still oiled, and your face resembles you at your best. In fact, I read an article the other day that said most women stick with the hairstyle they had at thirty-seven because that was their favorite period. Ladies, the hair knows. But as my friend Paul Williams says, "Times been worse, friends all gone, don't get crazy, life goes on and on." It does, and if you're lucky you get to go on with it; considering that the alternative sucks, what choice do you have?

But we boomers who watched the moon landing, who grew up with the pill, who invented MTV and ATMs and accepted STDs as a part of life, who have made more money faster than any generation in history, and through the birth of the computer (our invention!) have turned the world into a truly global one— we are not about to go down without a fight. Since we can't stop time, we have decided to deny it, at least verbally. And the amazing thing is, this whole concept of the shrinking years trickles down. An unmarried girl I know in her early thirties, who is starting to freak out, said to me, "You know, if fifty is the new thirty, then thirty must be the new twenty-four, which means I'm really an unmarried twenty-four-year-old, which is totally acceptable—not an unmarried thirty-year-old, which is totally terrifying." I'm not sure how she came up with thirty being twenty-four. But she wanted the extra decade to find a husband

and start a family. It just proves the logic is so fuzzy you can manipulate the numbers any way you want.

When I first heard the phrase "Fifty is the new thirty," it was like Christmas morning, winning the lottery, or getting an Oscar—maybe better, because even when those things happen you still face old age. I actually went out and bought a cashmere sweater with Tinker Bell on it to celebrate my newfound youth. And the scary thing is, I wore it.

I wasn't the only one; I bought it in a shop where they were flying off the shelves. Clothes are one of the definitive ways we can claim our true age; would a fifty-year-old wear a Disney character on her chest in broad daylight? The answer is that the only people who would do that are clinically insane fifty-year-olds, truly desperate ones, or anyone who is spending the day in the Magic Kingdom without access to a mirror. But not even Tinker Bell could keep middle age away forever. She couldn't do it for Peter Pan, and even in cashmere she could not do it for me. The day I realized this I gave the sweater to my fifteen-year-old, who threw it back at me in disgust. She wouldn't be caught dead in it. It was even too babyish for my six-year-old; she wanted Hannah Montana.

As fifty loomed closer and closer, there was nothing remotely similar to thirty taking place. In some ways it was much better than thirty, but in many ways it was scarier and harder to navigate. There was no question it was going to require a major attitude adjustment on my part, and if I was going to continue deluding myself into thinking I was getting younger, I would only spiral further away from who I really was and who I hoped I could become. Fifty is not the new thirty; it's nowhere near the new thirty. It's a period so full of changes, adjustments, and loss that the reconfiguration of your life and psyche almost becomes a full-time job.

One of the most confusing things about being fifty is that there is really no game plan in place. No matter what trail you end up taking, there is a very clear life map from twenty to fifty, one that starts getting drummed into you at about age two. So there are many years to ponder the possibilities and look forward to the highlights. In fact, much of a female's childhood is spent fantasizing about and play-acting all the fabulous things that will happen between twenty and fifty. Someday my prince will come . . . being a mommy . . . having a job like Mommy . . . playing house . . . And if life turns out remotely well, many of those things will happen. You grow up, you get out of school, you fall in love and out of it several times, you get your heart broken and mended and broken and mended until it resembles a crazy quilt. During this period you get a job, then you get better at your job, and eventually you get a better job. One day you finally find a guy who calls back, says what he means, and loves the fact all your soup cans have to be aligned, so you get married. You have a few kids. If you wait too long you may have some in a petri dish or a nice lady may carry them or you could take a trip to Beijing, but the possibilities are endless and a parent you become. You raise your kids. Then either you decide to stop working and raise them yourself or you want and need to have it all, so you hire a loving babysitter and do both. No one said it would be easy, but it's keeping you active. Since you are a woman and you have to prove the kids don't affect your productivity, you get better at your job and become jealous of the sitter who is spending her days with your kids. If in fact you avoid the mommy trap, chances are you are on the full-throttle career track. This will keep you very busy, very focused, and way ahead of the pack. Plus you will become very adept at answering the "Why don't you have children?" question. You can skirt brilliantly around it with "There is always China." And eventually you can hold your head high and say, "I just don't want them, and I'll bet you my 401(k) is twelve times the size of yours!" With luck you stay married. But if you fall into the half of people whose marriages collapse, chances are you can find someone else by forty-five.

But then—dum de dum dum dum—*fifty*. No matter how much Botox you get, things will start falling apart: some marriages end, some kids are gone, some jobs are terminated, most faces fall, and all boobs do. But no one bothered to fill us in on this. When we were little and playing with our Fisher-Price house, we never practiced moving out of it because we didn't need as much room without the kids anymore, or because we couldn't afford it when Prince Charming moved in with his paralegal and hid the assets. No one ever said, "You get married, have babies, have a wonderful career, then wake up with an empty nest, edged out of your job by downsizing—read ageism—not to mention at the same time your face will be falling, your bones will be getting brittle, and your vagina will be dry as the Sahara. Your husband may or may not hang around for this part. Or you may decide you are sick of him. Either way you could be single with an extra seat every week at the symphony. Chances are your parents will be sick or dead or really needy. Some of your friends will start dying as well, but if you take good care of yourself, you'll feel great, look good, and have plenty of energy, but not much to do unless you prepare for it in advance." No one gives women a game plan for a hearty last thirty years. They forgot to write that into the story, probably because they didn't want to have to do the illustrations.

The truth is, fifty ain't thirty no matter how you cook the books. It's fifty, and it arrives with more baggage than Paris Hilton on a press tour. However, the good news is that if you are prepared, it's not as bad as it sounds, and it's not your grandmother's fifty at all.

1

Fifty Years of Fifty

I think when the full horror of being fifty hits you,
you should stay home and have a good cry.
—ALAN BLEASDALE

When I was born my grandmother was fifty-two, the same age I am today. There is no question that if you compare the two of us, I look a good twenty years younger. So, going by the "fifty is the new thirty" philosophy, that would make me a lean and mean thirty to her flabby fifty. Except her fifty looked more like our seventy.

At fifty-two my grandmother had short gray hair kept tightly curled with a weekly visit to the beauty salon and a daily dose of Dippity-Do. Anything else would have been inappropriate and time-consuming. She wore glasses. Not fashion statement glasses, but a seeing aid. (When she turned sixty-five, these were joined by a hearing aid.) She could have had contacts, they existed in 1958, but knowing Grandma, her response was probably, "You put what where?" She could have had that printed on a T-shirt. ("A T-shirt? Like the baseball players?") It's lucky she didn't wear T-shirts, as she was, well, really fat by any decade's standards. I suppose she had an excuse; it was the fifties and there wasn't a health club on every block, though Jack LaLanne had sprinted onto the scene and opened his first gym in 1936. He was on TV every day from 1951 to 1985, doing sit-ups and push-ups and drinking Rocky-worthy elixirs. If people had paid attention to Jack back in the fifties, no doubt they would have had a leg up on aging. But he was a lone voice in a postwar world of prime rib, martinis, and cigarettes. America had suffered a lot and now they were going to have fun, and sit-ups followed by raw eggs sounded like anything but. It might not have been pleasurable, but it clearly worked. As I write this he is alive at ninety-six, still works out two hours a day, and apparently pulls tugboats with his teeth in his spare time. I met him once and he literally bench-pressed me.

But my grandma Dot was not listening to Jack's "lunatic ideas." "And what happens once I get on the floor? Is Mr. Tight Pants going to come over here and pick me up?" If she'd put down the cinnamon buns and spent a little time working on her own buns, she might not have needed a few Bufferins and an extra pair of hands in order to get up from the chair. But tight abs and butts were as appealing as putting little pieces of glass on her eyeballs. And besides, if God wanted you to have tight abs, why would he have invented the girdle? And boy oh boy, did she have a girdle. If you'd split the seams and hooked it up to a hose, you could have used it as a Slip 'n Slide. It ran from her knees to her bra. And her bra was so big you could fit an eight-pound baby in each cup. Then once she was sufficiently held together she covered herself in a blue dress—nothing fancy or tailored to highlight the good or camouflage the not so good, just a solid blue dress purchased at Lane Bryant, preferably on sale. At home she let it rip, took off her girdle, and put on a muumuu. Aloha!

Dot did not follow Jack's dietary tips either. Jack said, even back then, "If man made it, don't eat it" and "If it tastes good, spit it out." Grandma did exactly the opposite. Ketchup was her vegetable of choice. But what really kept her motor idling were carbs, the whiter the better. Potatoes, mashed or baked or hash browns—oh, did she love hash browns, and they went well with that nutritious side dish, ketchup. Rice was okay, and pasta was still the lowly spaghetti, though an opportunity to consume some vitamin-C-rich tomato sauce. She clearly knew this was not modeling the best dietary habits, so occasionally she would order peas, take a few bites, then push them my way.

"You need your vegetables."

"So do you," I would say, perhaps channeling Jack LaLanne.

"I'm allergic," she retorted as they made their way back to my side of the table. She also swore she was allergic to everything she didn't like, including all other vegetables. I'm not so sure this was true, though it gave her a lifetime hall pass on nutrition.

What she really loved were desserts. All desserts. She was an equal-opportunity dessert devourer. There are photos of her with her chums sitting at the Brown Derby deliriously

wolfing down hunks of coconut cake. She didn't stop at the coconut cake—there was the cherry pie at Du-par's, the rice pud-
ding at Ships, the meringue pie at the Beverly Hills Hotel. And
there was a cobbler at Hamburger Hamlet to die for. Remember,
this was every day, after every meal, not just a snack or a treat.
Plus there was always a midmorning snack "to keep my blood
sugar up."

The whole regime is doubly confounding to us considering
she was not on her way to work it off in a Bikram yoga class.
In fact, the afternoon snack was followed by the only cardio of
her life: pushing the market cart around the A&P, collecting as
much man-made food as she could.

The saturated fat of her diet somehow made its way into her
beauty routine—if you can call it that—in the form of Crisco.
I kid you not. "Face creams, shmace creams—those are for the
Rockefellers, and they don't work. All you need is a little Crisco."
And she'd dip her finger into a tin of Crisco and smear it all over
her face. Crisco being Crisco and Grandma's face being as ne-
glected as it was, the stuff would not glide on smoothly; rather, it
would get stuck in the crevices, giving her a kind of fright-mask
look. She would even try to use it on me, adding, "It's never too
early to start!"

There were two cans of Crisco in her house at all times: the
one for her face and the one for cooking. Emma, her house-
keeper for decades, did wonders with it in the kitchen. She was
the same age as Grandma but had no wrinkles, and I never saw
her put the Crisco anywhere near her face. The odd thing is, my

grandmother was not a poor woman from the sticks. She had money and lived in Beverly Hills. She could have afforded whatever top-of-the-line wonder products were in vogue in the early sixties. Helena Rubenstein, Elizabeth Arden, and Estée Lauder were in all the department stores back then. But we would breeze right by them on our way to the children's department or—sigh—the Tea Room for some triple-layer cake or a raspberry cream pie. And then in my teenage years, when I would make her stop at the cosmetics section, where we would restock my groaning makeup bag, she would never buy anything for herself. She would suggest Vaseline instead of Clinique moisturizer and insist witch hazel was just as good as Bonne Bell 1006 (nothing was as good as Bonne Bell 1006, except maybe Yardley Good Night lipstick). Despite my ignoring her suggestions, she'd happily pay for my indulgences, knowing she would soon be face-to-face with an éclair.

My grandmother was totally unapologetic and not at all remorseful about her age or physical deterioration. I never once heard her lament the lack of elasticity in her skin, the droopiness of her eyelids, or the craters of cellulite that made her legs look like a map of the moon. She never compared her face of today with the one of her youth; she didn't long for the days of thin limbs that didn't creak or her long blond curls. She would proudly show photos of a younger, leaner, more attractive version of herself, but it was for strictly familial, historical purposes. "This is when we spent the summer in Coronado; the one on the left with the smirk on her face is your aunt Yetta."

Grandma was a woman in her fifties; this is how they looked. "And I'd like that pie à la mode, if you please."

You have to remember that these were the women who lived through the Depression and World War II. The overpriced indulgences that have become such a part of our self-obsessed boomer lifestyle would have been quite appalling to most of them. They had survived enormous deprivation and sacrifice—twice, in the case of my grandmother. Her father lost everything in the Depression, which left an indelible scar and a lifelong reluctance to spoil herself in any way, and then she not only lived through the war but devoted her time to the people who were fighting it. She ran the blood bank in Los Angeles. She sat with wounded soldiers and read to them. When she came home at night she would write letters to those overseas, boys she didn't know but who she felt needed to know that someone cared. Like many women of her generation, women who were not in the workforce, who did not have careers or identities separate from being a wife and homemaker, she looked back on these years as the highlight of her life, a time when she was needed and had a purpose.

The memories of living through two such significantly difficult times clearly influenced not only my grandmother's perspective but also the way others of her generation saw themselves, as Americans and as women. The experience of going without and surviving made it impossible for some to indulge in the kind of frivolity and waste we feel entitled to today. It makes me wonder: if today's technology had been available then, would she have taken advantage of it? My gut says no. Not in her case. Her younger sister died of peritonitis as a child—she was part of the group for whom childbirth and childhood were still danger- ous and who had either witnessed or experienced the ravages of polio. They grew up without antibiotics; a bad flu could end it all (and, for many, did). I think all in all they were just so damn happy to have made it. Wouldn't it be asking too much to have beaten the life expectancy odds and have a creaseless forehead to boot?

I'm not so sure this attitude made her any happier. She was very high-strung and took Librium constantly. My guess is today she would have been a Prozac girl. But I don't think the unhappiness came from her lack of youthful, attention-grabbing looks. She was sad about her life in other ways, ways she kept hidden from herself and those around her. But she was thrilled to be a grandmother. And if looking like one went along with the job, fine with her. She was neither religious nor contempla- tive. She never got within twenty feet of a therapist's office or read a self-help book, but she was totally accepting of her station and appearance at that stage of her life. And because of that I was more comfortable around her than with anyone else on the planet. Granted, she spoiled me rotten and was the only person in my life who was thrilled to do anything and everything I wanted. And what kid doesn't want to hang out with someone whose idea of a meal is three desserts?

I once spent a summer with her and gained twenty pounds. Now, that may not sound like a lot, but I was eight at the time.

Those daily cobblers tend to add up. But I didn't care: I was roly-poly and happy and didn't want to go home. Once I did, my mother stuck me on a diet before you could say whipped cream. But I remember missing my grandma, and not just for the sugar rush. The pleasure she found in just being my grandmother was intoxicating and comforting. It was unconditional love—the kind of love that is impossible from those who are obsessed with how they look, how they used to look, or how they might look with just a little more Botox.

I don't think I would recommend Grandma's regime as a way of life, nor do I follow anything resembling it. But perhaps she had a version of the right idea. Maybe just accepting who you are, buying a bigger girdle, and celebrating your age with ice cream on the side of everything is the way to go. The sugar did ultimately cause her to suffer from diabetes in the last years of life. But it didn't slow her down or stop her from indulging. And the truth is, her lifestyle did not affect her longevity one bit. She died in her sleep at eighty-six. They had to write something on the death certificate, so it said "arterial sclerosis." In the end, it was death by Crisco.

My Mother at Fifty

I don't know if it was in defiance of her mother or just her own vanity, but my mother jumped on the antiaging bandwagon before they had tightened all the wheels. I remember her rubbing my cheek and commenting on the lack of wrinkles—when

I was eight! She would hold her hand up to mine and compare her bulging veins to my barely visible ones. I can honestly say I do not remember a time in my childhood when my mother was not actively pulling, rubbing, comparing, researching, buying, and applying any and all techniques that would somehow keep the bogeyman called age away from her door.

There is no question she was way ahead of her time. How she got there I cannot tell you. I remember finding the whole thing annoying, distasteful, and downright weird. My friends had bologna sandwiches on white bread for lunch, but I had artichokes. (Actually, I didn't have a lot of friends, as what really popular kid is going to eat next to a seven-year-old with an artichoke?) Mom didn't follow the diet of Jack LaLanne, as he was way too mainstream for her; she somehow found her way to Gayelord Hauser, who was said to keep Greta Garbo however young she might have looked. (It's a hard curve to grade on, considering she was a recluse from her mid-forties on.) Hauser was the nutritionist to the stars and, much like Jack LaLanne, was exposing the diabolical effects of sugar and white flour just as the postwar industrialized world was discovering those products and embracing them with a vigor that continues to this day.

Thanks to Gayelord I was not allowed much sugar as a child, and I remember vowing to feed my kids only candy once I became a mother myself. Instead I was forced to drink this third-world cabbage soup my mother assured me would keep me young forever, totally oblivious to the fact that when you are eight and counting your aging process by quarters of years, perpetual youth is the last thing on your mind. We would slurp up the smelly soup, then she would run to the mirror to see if it was working. She spent a lot of time pulling up the sides of her face to see how much better she would look with a tightened visage. And she was only in her thirties at this point. Women twice her age were not behaving this way yet. It was decades before twenty-year-olds would freeze their foreheads into eternal smoothness.

But aging was something she was going to run from as fast as she could, and there was no limit to the places she would go, the cures she would search out, and the products she would buy.

While her own mother was spreading Crisco on her face, my mother was a devotee of Erno Laszlo. Estée Lauder, Helena Rubenstein, and Elizabeth Arden made high-end, beautifully packaged facial products, but Erno Laszlo was a way of life. His products were known as "the jet-setters' choice." Erno Laszlo was to face cream what Gayelord Hauser was to food—a revolutionary. "Seventy-five years ago he developed the first dermatological skincare line based on five basic principles. Including splashing, teaching the critical importance of clocking and the daily ritual." The critical importance of clocking? It sounds ludicrous even by today's standards, so imagine it forty years ago. Clocking was what you did to judge a race, a standardized test. It was 1968, and nobody clocked their faces; they barely moisturized them. But Mom figured if clocking was what it would take to look younger, then clocking it would be.

The foundation of Laszlo was a tarry mud soap that you lathered up and followed by thirty rinses with scalding hot water—*exactly* thirty, not twenty-nine, not thirty-three. (Remember, ritual and clocking were the foundational principles of his regime. Those who deviated might not be able to externalize their inner Garbo.) He also advocated steaming, so when Mom wasn't rinsing, she had her head over a pot of boiling water. Mrs. Stoltz down the street used boiling water for macaroni and cheese (pure poison in Hauser's book) and Pond's cold cream on her face. I liked Mrs. Stoltz a lot. I preferred macaroni to artichokes, and I loved the smell of Pond's.

But my mom devoted a half hour in the morning and a half hour at night to the clocking, the mud soap, and the boiling water. A half hour? That was an entire episode of *Green Acres*. But what was a missed sitcom and a faceful of scalding steam in the name of eternal youth? The soap ritual was followed by the lotion ritual, which because of that omnipresent clocking was not the same at night as in the morning. See, the clocking was a function of the fact that our faces are drier and oilier at various times of the day, so we should attend to them accordingly.

The product I remember the most was the Active pHelityl Oil. According to the directions, you were supposed to put it on before the soap, but Mom, being a bit of a beauty maverick, decided it was better if she smeared it all over her face and left it there. I guess, being Grandma's daughter, she didn't fall too far from the Crisco tree. Her decision to let the oil stay meant she was always kind of shiny and glossy, which I think in her mind and mirror looked dewy, which translated into young and fabulous. I remember it looking like she'd forgotten to wash her face, which was ironic, considering the twice-daily splashing ritual. But the most annoying part of the whole thing was that she left a thin layer of oil on anything that got within inches of her face, and oftentimes that was me. If you ever used the phone in our house, you walked away with a visible film of Active pHelityl Oil that required thirty rinsings to remove. Perhaps many of her friends look younger than they should today because of it. (Though I must hand it to her—as a devotee, she kicks butt. She's eighty-one and has never stopped clocking and rinsing, and she and her belongings are still covered in a film of Active pHelityl Oil. Truth be told, she looks good for her age.)

Mom being as determined as she was, the jet-setters' products and Greta Garbo's diet doctor were not enough for her. There were more secrets out there, additional keys to the kingdom of eternal youth, and after her hour a day of Laszlo, she was left with twenty-three hours to find them.

Somehow, somewhere in the mid-sixties she learned that the sun was bad for you. This posed a problem for her, first because we lived in southern California, but mostly because she was a lone traveler on this road back then. Sunscreen as we know it did not come on the market until 1972; even then it was just an SPF rating that, much like the label on cigarettes, was ignored by many, and there were very few products that even contained it. Plus the massive literature touting the dastardly effects of the sun was a few decades away from public view. The first time anything resembling sunblock was used was during World War II, by soldiers who were exposed to too much sun in the Pacific. Apparently the stuff was red and sticky and not particularly effective. The actual concept of sun protection factor, SPF, was developed in 1962, but it was in Europe and, like many skin care products, did not make its way to America for many years.

The late sixties were still the days when you couldn't be too thin, too rich, or too tan. Coppertone (remember the cutie with her white behind?) and Sea and Ski were what most people used; the jet-setters had that divine-smelling, transparent orange gel Bain de Soleil, but it was expensive, and this was also before entitle-itis hit our shores. You bought what was within your budget. If you were really hard-core, like us California girls, you brought out the baby oil, the cocoa butter, and the reflectors. The last thing tanning was associated with was wrinkles—in fact, a tan made you look younger. You'd think with this being the case, my mother would have been a glossy, nutty, Active pHelityl Oil–covered brown.

But once again, way ahead of the pack, my mother somehow learned about sunscreen and that it would prevent wrinkles. How she figured this out I have no idea. I cannot emphasize enough how aging was not at the center of people's conscious-ness; even the movie stars whom people looked to for their beauty guidelines were tan, tan, tan. So average American moms, especially in California, were tan; they golfed and gardened and sat in the sun. They looked cool and smelled like Sea and Ski. But my mom was not average by anyone's terms. Somehow, some way, she found her way to the earliest sunscreen. This was so early she had to order it from Switzerland. And she had to order a lot at a time, so huge boxes would arrive packed with these little black and white tubes.

In retrospect the formula was probably closer to what the soldiers used. While it wasn't red and sticky, it was brown and gooey, the consistency of very heavy makeup base, and the skin never really absorbed it. But that didn't stop her from spreading it all over her body and mine whenever I let her close enough to do it. And because she had so much of the stuff and was convinced it was a miracle in a tube, the tubes were everywhere. There was always one on the front seat of the car, the lid never quite tight enough, oozing its chalky contents onto the seats and whatever got in its way. Mom would slather it on her hands, as she decided driving made age spots appear. "Look at my left hand, Tracey, the one always near the window. It looks ten years older than my right." She would then squirt out a dollop of sunscreen and smear it on her hand, getting a third of it on the steering wheel.

She kept a half-open tube in every bag, so her purse and her belongings were always coated with some remnant of the cream. Weirdly, this didn't bother her; she was much more fearful of being without it. And she didn't run around telling people about it, either; it was her personal secret. Everyone else would look like an old Indian chief and she would look like Liz Taylor in *National Velvet*.

When we went to the beach my goal was to fit in with the kids, the same ones who got bologna when I was lunching on artichokes and Gayelord Hauser broth. They, of course, were smearing on cocoa butter and frying themselves to a California crisp. I would spread out my towel and reach for my hidden stash of cocoa butter, and within seconds there would be Mom running after me with a tube oozing the brown gunk. She would literally chase me down the beach, and since, sad to say, youth is something that does not come in a tube, I was faster. (Of course, the fact she was wearing a full-length caftan, her normal beachwear for years, didn't add to her middle-age speed).

"Get back here right now, Tracey Dee Jackson, if you know what's good for you!"

"Tanning is good for me!"

"Tanning will make you old and wrinkled—you want to look like Nana Dottie?" (My Palm Springs–based, perpetually tan grandmother, and her ex-mother-in-law.)

"She's seventy years old!"

"Well, she looks ninety, and that's because of the sun . . . well, that and her nasty personality."

And on it would go until one of us gave up. Usually it was me, as those were back in the days when parents actually had some real authority. If she wasn't staying at the beach, she would get me with the stuff en route. At some point in the car ride, she would grab one of the ever-present tubes, squirt the gunk on, and yell, "Rub!" Not that rubbing it made a difference; it still just sat beige and pasty on the first layer of your skin.

"On your face!"

I sat motionless.

"If you want to go to the beach, then I suggest you cover your face—*now*."

I reluctantly rubbed the damn stuff on my face as I listened to that motherly triumph: "You'll thank me someday."

"In your dreams," I would mutter.

But okay, here it goes—publicly: thank you. I thank her all the time. Not to her face, of course. But when the facialist tells me I have nice skin, even after she's been tipped, I say proudly, "I'm fifty."

"You don't look it—you have the skin of someone much younger."

"Well, my mom put sunscreen on me as a child."

"Lucky you, and smart mom." History and my face have proven those things to be true.

But Mom, who must have had some international pre-Google system of unearthing hidden global secrets to eternal youth, was not anywhere near satisfied with the results from sunscreen, cabbage soup, and her clocking ritual. She wanted more—more immediate results, more intense and visible signs she wasn't aging. I think she wanted to look like me and couldn't figure out why I looked so young. I kept reminding her I wasn't even thirteen yet. I was conflicted: my inner fashionista (soon to be a full-fledged trend follower) thought my mother looked glamorous. She had hip clothes, Roger Vivier shoes, and a Louis Vuitton bag, albeit with a bottom and handles splotched with sunscreen, and she did always look a little oily. The adolescent me who was dying to fit in with the pack thought Mrs. Stoltz looked better in her golf shorts, Keds, and tan. But in the end the inner fashionista won out, though I still think Mrs. Stoltz looked like a mom should look.

From the thirties up to the eighties, Eastern Europeans were at the forefront of all skin care, diet, and health discoveries. With the exception of Elizabeth Arden, who came from Canada, of all places, and Jack LaLanne, who was American born and bred, the others—from Hauser to Lauder—were either Czechs, Hungarians, or Romanians. Which does not fit in with the babushka-wearing, apple-faced images we have of prewar or midcentury Eastern Europeans. People like to give the French credit, but it was scents and then the application of it all that they excelled at. The ones in the labs making the discoveries, trying to figure out how to retard the aging process, were further east, and by the time my mother discovered them, they were behind the Iron Curtain.

That my mother was always on the cutting edge was no doubt partly due to the fact that she read *W* and *Vogue* as if they were the cornerstones of a PhD thesis. Both could always be counted on to enlighten readers about the latest antiaging fad celebrities were hooked on at the moment. Again, in those days it was primarily celebrities who were doing these things. They had not trickled down to the upper middle classes, much less the middle class.

Somewhere in her research my mother discovered Ana Aslan,

a Romanian biologist and physician who is credited with being the first person to latch on to the theory that the aging process could actually be retarded. She was also a master of contradictions: on the one hand, she said, "To be young forever doesn't mean to be twenty years old. It means to be optimistic, to feel good, to have an ideal to fight for and to achieve it." Yet while she was saying this she was injecting celebrities such as Charlie Chaplin, JFK, Marlene Dietrich, Kirk Douglas, Charles de Gaulle, and even Mao Zedong with her fountain-of-youth potion, Gerovital H3. If you go online today you can buy the stuff with a click of your mouse, though it still has not been approved by the FDA. But back in the early seventies there was only one way to get it, and that was to go to Dr. Aslan herself. The only problem was that her clinic was in Transylvania.

But neither the Cold War, a basically closed-off Communist country, nor mountains of red tape were going to keep my mother from the real-life Shangri-la, even if it happened to be Dracula's hometown.

This was the 1970s, and people were terrified of that part of the world, but before you could say "saggy jowls," we were on a forty-eight-hour Pan Am flight that made endless stops to refuel and eventually landed in Bucharest. I was a good sport as a kid, and by this point I was used to my mother's crazy schemes. But this one was very hard for me to wrap my head around.

"Transylvania is where Dracula lives! Is it going to be like *Dark Shadows*?"

"Better. They are going to make Mommy look younger."

"I thought the sunblock and the yucky soup did that."

"Yes, but this is going to make Mommy look really young, maybe as young as you!"

Try to process that when you're twelve. First off, you can't imagine why anyone would want to look like you. I wanted to look like Marlo Thomas in *That Girl*. If she had said we were going to travel for two days to Danny Thomas' house so Mommy could look like Marlo, that I might have understood. But going to Dracula's house seemed a little ominous and scary, especially since it was being sold to me as a spa.

It was the longest trip I had ever taken, and I had no idea what to expect. I knew nothing of Cold War Europe, but clearly Mom knew it might be tough going, because she kept promising me I would love it. "It will be fun! There's even a swimming pool on the twenty-third floor of the hotel!"

As we disembarked from the aircraft we were met by rows of soldiers with their rifles pointed at us. "They don't do this when we land in Honolulu," I said, clutching her hand. The Romanian soldiers stared at me. I was missing *Dark Shadows* for this?

"This is going to be fun! We're going to have an adventure, and Mommy is going to look ten years younger when we get on the plane home next week." I think she was trying to convince herself as much as me at the moment.

My mother had made meticulous plans: we would spend three days in Bucharest, taking in the city and enjoying the five-star Intercontinental Hotel, with me splashing away in the pool on the twenty-third floor. But we drove through streets that had a gray pallor unlike anything I had ever seen. It was cold and bleak. I was hoping the pool was indoors. The people were gray too and looked much older than the people in Santa Barbara; so far it did not seem like this was the place to come for eternal youth. Rather, it appeared to be the place to come for eternal depression. We got to the Intercontinental, which had been sold to me as "like the Beverly Hills Hilton." It too was gray, not shiny white like the Beverly Hills Hilton. It might have been white when it was built, but by the time we got there it was covered with a layer of gray soot from the coal burned everywhere in the country.

My mother was not deterred by the overwhelming grayness of it all; she was on a mission. She strode up to the front desk of the hotel and plunked down her credit card, all her documents, and—most important—her confirmation letter from the Aslan Clinic for her prepaid-in-full youth treatments. The reservation clerk looked it all over, slowly and suspiciously and repeatedly; one paper, another paper, the passports, the visas, the bus tickets to Transylvania (also prepaid), the Aslan Clinic folder and papers. Eventually he placed them into two neat piles, our personal travel documents and tickets in one pile and the papers from the Aslan Clinic in the other. He pushed the Aslan documents toward her and announced, "Not possible."

"What's not possible?" she almost shrieked.

"National holiday, no traveling outside of Bucharest allowed, government policies."

"You must be mistaken; I've paid for it," she insisted, Western commerce being the only policy she understood.

"National holiday, no inner land travel, you stay in Bucharest."

"No, we're going to Transylvania," she said with great authority—he didn't scare her, or if he did, he wasn't nearly as frightening as the thought of a face marred by wrinkles. And she certainly did not understand what totalitarian government really meant.

"Why you are wanting to go to Transylvania?"

"We are going to get the youth treatments," she announced.

He looked at her, then down at me, and back to her. I'm sure that compared to his wife and the other women he was used to, my sunscreen-protected, Active pHelityl–shiny forty-year-old mother looked like Marilyn Monroe. Why on earth was she in this part of the world seeking youth? And who was the midget with her? It had to be a midget; no twelve-year-old traveled into Cold War Romania seeking youth. No twelve-year-old sought youth, period. Even Americans weren't that crazy. He'd probably seen enough black-market James Bond films to convince himself that people who looked like us came to Romania in search of secrets, not wrinkle creams. So clearly Mom was Ursula Andress, who would seduce some sad sack into divulging a government plot that would not only bring down the country but cause the hotel clerk to lose his job and his sausage rations. And I was her midget spy whose hands were small enough to reach into whatever I had to reach into and grab whatever it was we needed in order to accomplish this nefarious goal. You could see the wheels of his programmed brain turning as he imagined Mom cavorting with the horny head of some government division while I hid in his desk drawer to snatch the submarine blueprints. As far as he was concerned, we were Captain Evil and Mini-Me, and his life depended on our not going to Transylvania.

Clutching her now useless Ana Aslan coupons and bus tickets, my mother started to cry. "Now I'll never be young."

"It's all right, Mom, you look really young."

"And now we're going to be stuck here for ten days." She was still weepy as we entered the elevator.

"But Mom, at least we have the pool on the twenty-third floor." I looked over to the panel of buttons to punctuate my cheery point. The hotel stopped at the nineteenth floor.

Since the Transylvania scheme had been a total bust, she needed a new plan for her attack on aging.

We had been back from Romania for about four months when she said she would have to go into the hospital for "a little procedure." I didn't really think much of this, as she was plagued with gynecological issues, which in those days required hospitalization instead of a visit to a doctor's office. It was fine with me, as that meant a few days with Grandma, and the Gayelord Hauser meals would be replaced with Baskin-Robbins, cheeseburgers, and lots of toys.

Three days later I walked back into the house. (Grandma must have known I would need some fortification, so we got extra cakes to bring home, and Barbie's Dream House just because.) I heard a garbled voice from my mother's room: "Honlle-diiddooommeee!" Had she learned a new language while she was gone?

I went into her room, and despite the fact that this happened forty years ago, I can see it, smell it, and feel the chills run down my spine like it was yesterday. Sitting in my mother's bed, wearing a bed jacket (a sort of see-through bed sweater that was considered chic or sexy or important back then), was a monster. Her head was wrapped in so many yards of blood-stained bandages, it was probably five times its normal size. Her eyes were little slits, which appeared to be held on to her face with black stitches; the part of her face that wasn't black and blue was red and puffy. Her mouth could barely open, so the words could not get out and nothing but liquid could get in. But the coup de grâce was the two long tubes filled with flowing blood that protruded from the top of her head. It looked like the alien from *My Favorite Martian* had taken a trip to the Twilight Zone.

"Moommmezzzzaadddaafaaaceeethift," she managed to gurgle.

I was too terrified to move. I had never, ever seen anything like this, not even on *Dark Shadows*.

"Another one of your mother's crazy ideas," Grandma muttered. "She had a face-lift—she wants to look young."

"Ithgooingtooooobebuuutifulll!" She couldn't smile, but the bloody tubes wiggled in delight.

"She thinks she's a Rockefeller or Elizabeth Taylor," Grandma said mockingly.

I looked at Grandma and her wrinkly face, which she could actually move. I looked at Mom with the tubes of blood and the slits and the bandages and the stitches. I vowed then and there I would never ever, no matter what, get a face-lift. (This despite the fact that as the years progressed, I could see that my mother looked much younger and was definitely less wrinkled than her face-lift-less contemporaries.)

If you look at my mother today and compare her to her mother at the same age, it's clear that the soup, the clocking, and the face-lifts have had a profound effect. She looks great for her age. She has relatively few wrinkles. Her eyes look better than mine, as she does not have the dark circles and bags I inherited from my father. (They say one eye lift in your fifties or sixties will last a good fifteen to twenty years.) And she has a vitality my grandmother lacked. The vitality has everything to do with the fact that, unlike Grandma, my mother started working in her late thirties and still does to this day. She writes books, lectures, and has a full life. She does not seem like an elderly woman. When you put these two women side by side, they are an ad for the war against aging.

Me at Fifty

I turned fifty a couple of years ago. I'm not remotely like my grandmother, with the who-gives-a-damn, let-it-all-go attitude. Nor am I part of the group that had to blaze a trail, like my mother in search of youth cures. I'm part of the injectables-dermabrasions-minilifts generation, the life-is-a-buffet-table-of-antiaging-solutions generation. In the 1960s women my age watched Ed Sullivan; this crop can turn on *Nip/Tuck*.

There is no question that thanks to my mother's early pioneering, good genes from both my parents, and the time in which I was born, I look younger than both my mother and grandmother when they were the age I am now. Fifty has changed in that way. If you do a lot of legwork in your thirties and forties, you will get to fifty looking a little younger than the generations before you. But it doesn't happen on its own. Go take a walk through a mall and you'll see overweight, unhealthy women who look three times their true age, or at least much older than their more diligent counterparts.

I would have to say that the biggest difference between my grandmother, my mother, and me is that I exercise and have done so all my life. Grandma simply refused to, and in her defense, it was not part of her world. Unless you were truly cutting-edge or attached to a sport, the yoga/Pilates/spinning/stepping/jogging route to youth was not part of the language.

Strangely enough, my mother, for all her searching and yearning, not to mention her globe-trotting antiaging tactics, avoided the one thing that in the end might have done her the most good. She would tell you she swam, and in fact she did swim. In the evenings, when the "treacherous" sun was gone, she would swim very slow, ladylike laps with her head held high above the water, where she could carry on a conversation, never missing a beat for lack of breath. Boomer exercise it was not. But in the same way my grandmother pushed the peas onto my plate knowing they were good for you, my mother knew exercise was important; she just couldn't be bothered. So I would do it for all of us.

We are all naturally chubby. We come from a long line of hearty Jews, and my mother, despite Gayelord Hauser and the artichokes, gave into her genetic adoration of sweets and simple carbs at a certain point and got heavy and became diabetic like her mother. It's a shame, as she had such a leg up on everyone in the youth department.

But as long as I can remember, she was pushing me to exercise: tennis camp, modern dance classes, more tennis camp, swimming where I actually had to put my head underwater and could not talk at the same time.

And then when I got older she used those same whatever-the-celebrities-are-doing-must-be-good-for-you tactics to find me the newest, hippest forms of exercise. When I was twenty she read about a guy in LA called Ron Fletcher. He was a disciple of Pilates long before anyone other than dancers was a convert to the method. He opened a studio in Beverly Hills, which after a few well-placed articles was littered with celebrities. An article from *W* or *Vogue*, still her bibles, would arrive in the mail from her with a Post-it attached: "Try it." And off I would go to find myself on weird machines working out next to Candice Bergen.

Six months later she would tell me about *the* place to go—Bikram yoga. "Everyone is doing it."

"I thought they were all doing Pilates."

"They've moved on," she would say. "And besides, yogis live to be a hundred."

In 1978 there was not a yoga mat in every backseat; it was a weird hippie thing the Beatles did. But in Beverly Hills an Indian man called Bikram who someday would become very famous was ensconced in a basement underneath the Fiorucci store.

The room was tiny and 110 degrees. Bikram was small but possessed the self-assurance that comes from being certain you know something that few others do. He would strut around yelling at Quincy Jones and Peggy Lipton, who would stop mid-cobra to kiss, and at Shirley MacLaine, who was just beginning the journey into her past, and at Candice Bergen. (Wait, wasn't she just at Ron Fletcher's?)

So off and on for years, even after he moved up and became well known and took over bigger spaces, I took Bikram yoga. I also did Radu, Lotte Berk, and—most fun of all—Richard Simmons when he was so new to the game he collected the money as you walked in and put it in a piggy bank. This was several years before he became a household name and the good-natured butt of many gay jokes. For whatever it was worth (and I believe it was worth a lot), my mother got me hooked on working out. And despite the fact that I have workout ADD, I have always done *something* on a regular basis. I never did it to look young; it just became a habit early on, and I am terrified of being fat.

When I first met my husband he was overweight and thought the fact that he would sit on an Exercycle drinking single-malt Scotch and reading Henry James while a baseball game played on the TV in background meant he was exercising. I told him to get to a gym fast. I read about a new one two blocks from our apartment. By then I had moved on to something called Core Fusion, which I knew would not be his thing. So he reluctantly signed up for a few sessions at the gym down the street and came back glowing after the first day. He had tapped into the endorphin rush, he loved the trainers, and the coolest thing was that Candice Bergen had been on the treadmill next to him. Once he told me that, I said, "No doubt you're in the right place."

He's been working out there for ten years six days a week ever since.

I can honestly say I was never really focused on looking young until I realized I didn't look young anymore. Just as I remember my first sight of my mother post-face-lift, I remember the first time I was aware I looked old, older than the image I had of myself for decades. I was forty-five at the time. I had no idea of what really lay ahead of me. The sagging jowls were nothing compared to what it was going to be like to go through menopause, lose my libido, lose my jobs to younger people, and lose my friends to disease.

All around me people were yelling I was thirty. We were fifty, but we were really thirty. When my grandmother turned fifty, she just took it on and lived it like she was told to. My mother fought it like it was a wildfire about to engulf her in flames. And me? Like many of my generation, I never really thought it would happen, and chose to deny its existence. How I've handled it and continue to handle it is the story of this book. It's my journey to fifty and beyond. It's not my grandmother's fifty. It's not my mother's fifty. It's my fifty, and I can promise you one thing: it is not remotely like thirty.

2

Menopause or Menostop?

You never realize how much you miss something until you lose it.
—EVERYBODY WHO HAS EVER LOST SOMETHING THEY
DIDN'T APPRECIATE BEFORE IT WAS GONE

At thirty the last thing on my mind was menopause. At fifty I couldn't stop thinking about it. At thirty I was concerned with finding a presentable straight guy who had as few personality disorders as possible so I could get married, write thank-you notes, have a baby, and write more thank-you notes, at which point my life would begin and I would write a big thank-you note to God.

I got my period every month on time—unless I got pregnant, which I did very easily. Let's put it this way: *you* had sex, *I* got pregnant. I had what I now know was a lot of estrogen. I didn't have a clue that I had an abundance of it, nor did I know what it meant. I didn't learn any of this until I woke up one day and it was gone—*forever*. There is no "pause" in menopause. It should really be called Menocometoascreechinghaltyouwillneverfeel-thesamewayagain. That is not entirely fair; it peters out in a condition they refer to as perimenopause, which can start as early as the mid-thirties.

There are endless books out there you can read about menopause, estrogen, progesterone, and all the "ens" that have been making your engine run your entire life. You can Google estrogen and get the molecular compounds and all the science-medspeak you need. I will give you the crib notes as I understand them. Estrogen is the hormone running through your body that pretty much keeps you ready and able to have babies. You need it to get pregnant, and it keeps your skin looking young and wrinkle-free. It's first cousin to that other "en," collagen. Collagen also keeps Grandma Gravity from tugging on your jowls and connecting them to your shoulders. Estrogen keeps your vagina moist and supple so you can make those babies. It keeps you in a fairly good mood so you don't constantly piss off the people who may be attracted to you, so you can hook up with one of them and make some more of those babies.

You know those days before your period when you are homicidal, suicidal, or just plain grumpy and everyone from the butcher to your boss helpfully points out that "it must be that time of the month"? Well, it's the time of the month when your estrogen and your progesterone are out of whack. They're not in sync and neither are you. Then you get your period and you are suddenly smiley and normal again—that's when the dynamic duo are back in action. When it comes to your moods, estrogen works as sort of a continual chardonnay drip: it gives you a light buzz, takes the edge off. It's the better you. When it comes to your vagina, it's sort of like having Brad Pitt strutting naked in front of you at all times. It has that kind of effect. The estrogen you never appreciate until it's gone is there not only to make you attractive to the opposite sex but also to make the opposite sex attractive to you. It's what turns "no, no, no" into "yes, yes, yes." (When you're fifty, no tends to mean no, or "I'll give you a hand job . . . after *Grey's Anatomy*.")

And then, just when you really need it, Mother Nature pulls the chardonnay drip that allowed you to keep it all together, and you find yourself more depressed than you ever imagined possible. You find yourself crying because your lipstick rolled under the front seat of the car, and then you are screaming at the waiter because the salmon you need for your omega–3s is from a farm and not the wilds of Alaska. Losing the remote—or, worse, pushing the wrong button so the screen ends up all fuzzy and no matter what you do, you can't get the picture back—can send you into a total mental meltdown.

This is what actually happens. No one prepares you for it, and so you think you are losing your mind along with every other part of yourself you have known for the last forty years. When your mind seems like it's going in mid-conversation and you can't remember Nixon's first name, it feels like the next stop is the assisted-living home. "You know, the president . . . I mean the one who flew in the chopper, he lied . . . you know . . . shit . . . I miss my corgi, his name was Spikey." And then you start to cry and next thing you know, you're sweating buckets. ("Is it hot in here or is it just me?") I ask you: is there anything less appealing than a wrinkled, dry, emotionally out of whack woman who can't remember Nixon's name, spontaneously bursts into tears at the thought of her childhood dog, and is perpetually overheated?

Yes, indeed, along with everything else, estrogen is also your own personal thermostat. The jokes about "Is it hot in here?"—they are not jokes. You are hot in a way that feels like you're on fire from the inside, and it's usually followed by an arctic chill; trust me, you can see Russia from that chill. And then get ready, because another hot flash is on the way. Okay, this isn't everybody, I know women who say they never suffered for even a moment. They breezed through menopause without, well, a pause. I find this very hard to believe. I put these women in the same category with those who thought the pain of childbirth was a glorious experience. I think they should all go gather in an empty stadium or ladies' room (depending on how many in the world there actually are) and exchange bliss stories. Because for most people it's hell, and if they're honest, they will tell you that.

Of course, the one break you might get from these sweat-drenched days is a good night's sleep. But who knew the estrogen/progesterone combo is also nature's Ambien? Suddenly you can't sleep, or you can get to sleep but you wake up every hour on the hour—1:43, you're hot, throw off covers; 2:44, you're freezing, where's the quilt?; 3:45, oh God, hot again, covers off—and so you end up with about three hours of totally interrupted sleep. Night sweats seem to be the thing you hear about most with menopause. Women tell stories of waking up in the middle of the night with drenched sheets and hair that looks like it just came out of the shower.

There are also some other bonus miles you get without your estrogen: you gain weight. Or so they say; I have worked over-time not to. Though no matter what you do, it's hard to entirely eliminate the "muffin top"—that little bit of flab that hangs over your pants. It can be a muffin top or a triple layer cake; that depends on you. But if you hold on to the bad habits of your youth, you will find yourself three sizes bigger before you can say, "Do these come with an elastic waist?"

Is this sounding remotely like thirty to you?

I personally suffered everything except the night sweats—although I made up for that with constant, excruciating head-aches. I would get hot but not sweaty. (I must have some kind of genetically built-in antiperspirant: I can go to the gym, work out, and then get dressed again without needing to take a shower.) And it never let up. It seemed to get worse each day. For some reason the chills did me in. I would get freezing cold, shaking, then really hot, then nauseous, then I would feel like I was going to pass out, and finally I'd start wondering if ten Tylenol a day is a safe dosage. There is no question the quality of my life was being greatly compromised.

And the kicker is, there is a cure. There is a little pill, or a small patch, or a clear gel you rub on your arm, and it all mag-ically goes away or mostly away. This is HRT, aka hormone re-placement therapy. Some people think it's God's gift to women; others feel you might as well swallow cyanide. The only experience I had with HRT was that when I was growing up every woman I knew took it. My mother had a hysterectomy at forty and has been taking it for forty years. (I will add she has not gotten breast cancer or anything else they mention on the box the stuff comes in or on the news.) She went off it briefly at the age of seventy-eight and after one week went right back, insisting that if it shortened her life at this point so be it, she could not and would not live without it. My stepmother has been on it for years as well. She recently went off it herself for a miserable week, at which point she went right back on, citing the same reasons as my mother. This may be the only thing I have ever heard them agree on: life is not worth living without their HRT.

For decades millions and millions of women the world over were spared the humiliating, misery-making, life-ruining symptoms of menopause. They took their HRT and on with it they went. In fact, when I was growing up I don't remember hearing anyone complain about menopause or its symptoms, as most everyone was on some form of HRT. But much the same way AIDS put a damper on the sexual revolution, the Women's Health Initiative study and a few other published reports transformed the menopause-killer into the female-killer overnight. It went from the Mother Teresa of meds to the Boston Strangler.

By the time I hit menopause HRT had truly become the Chernobyl of medicine. Studies fingered it as the culprit for an increased risk in breast cancer, ovarian cancer, stroke, heart disease, and other life-threatening conditions, several of which—

like heart disease—we had always believed it was supposed to prevent. Woman everywhere threw out their pills, bought a bunch of fans and lubricants, and, according to one of my doctors, started taking antidepressants in record numbers to combat the mood swings. Same with sleeping pills, another pill that replaced the estrogen—though it was not in fact replacing estrogen but merely trying to alleviate some of the symptoms of not having any. Those reports changed the minds of many women and the quality of their lives forever.

Now, I am not a doctor or an authority of any sort. I don't claim to know if HRT is good or bad for you. Doctors themselves don't even like to do that. The trend now is that they give you the facts and then the choice is yours. They don't want to tell you to take it and have you walk in with breast cancer two years later and hold them responsible, despite the fact you might have gotten it even if you hadn't taken the HRT. They are covering their asses, and who can blame them? Some refuse to dispense HRT at all; others will open a drawer and it's packed with samples of the little suckers.

I personally don't recommend taking it, nor do I recommend not taking it. I am merely sharing with you my experiences and the choices I made that were right for me.

When I hit menopause I took the stance of many around me: I was not going near HRT, because it could kill you. I'm hypochondriacal enough; I need to take a pill every day that will increase my chances of breast cancer? I'll just suffer. I'm Jewish. I will suffer and complain bitterly. I will ruin my life and torture those around me. What else is a middle-aged Jewish woman whose body temperature is 104 degrees supposed to do?

I would powwow with my other menopausal friends, exchanging symptoms, horror stories, and tips. When we were in our thirties we'd talk about babies, preschool, our jobs, and whether our husbands ever drank our breast milk. Now suddenly it was "If you take soy with the black cohosh, you only sweat at night. Try acupuncture for the joint pain, ginkgo for the memory. If you don't watch TV or read in the bedroom and drink warm milk before bed, then maybe, *maybe* you can get in at least two hours' sleep. Don't forget to sprinkle lavender on your pillow. For the day sweats dress in layers, bring an extra set of clothes, and always sit near the AC in restaurants. You can use your saliva as a lubricant. I've heard SAM-e works on depression. The thinning hair, try Rogaine, it worked on my uncle Saul. There is a type of fan that is very quiet and rotates for the bedroom; you stay cool but your mate doesn't get hypothermia."

In the land of HRT there are two very distinct camps: those who take it and those who don't. The discussions above were clearly had with the latter, the slogan above their clubhouse being, "You might as well put a gun to your head if you take the stuff." I was a card-carrying member of that group—for a while. I'm not pointing fingers, and there are exceptions, but for the most part the women who pitched their tents in the anti-HRT camp were more out of shape and had let themselves go both gray and saggy, not to mention flabby. Their theme song was that life goes on, I've earned my wrinkles, and if God didn't want you to gain weight he would not have invented Spanx and Eileen Fisher. Then there were my friends who took HRT. They looked better, were less cranky by far, lubricated on their own, slept better, and had thicker hair, better memories, more energy, and tight abs (if they went to the gym); they swore by the stuff. Most told me it saved their lives and they would never give it up.

I was truly stuck between a rock and a hot place. I did not know what to do. All the health food store treatments were not working. I was crying daily and yelling at everyone, and I felt like hell. I started pondering the other side's position: should I play Russian roulette and take HRT? My mom is okay, and they say a lot of it is genetic. Or are all those studies right? Then I would visit Dr. MacBook—I would go on every website and read obscure studies done in Sweden on small groups and big studies done in Canada, most of the time not understanding what they were really saying and just trying to decipher the charts. My only concern was, if I take the stuff and feel like myself, will I get breast cancer? It was a very simple question, but no one was willing to answer it.

My ob-gyn had given me a sample box of the little pills. They sat in my medicine cabinet for months. Sometimes I would take them out and just stare at them, hoping if I stared long enough they would miraculously spell out the answer themselves: "WE DON'T CAUSE CANCER." But then I would unfold the warning pamphlet tucked into the box that would say in real letters exactly the opposite. So back into the cabinet they went, untouched—and I would take a cold shower and complain some more.

But as those of you who complain know, it does nothing for any situation except make it worse, and my husband was bearing the brunt of it. Men may not go through the symptoms, but they are the recipients of much of the fallout. He started asking me to get some form of help.

I was slowly, slowly moving in a direction, though I wasn't exactly sure what that direction was. The first step was that I wrote the following email to my ob-gyn:

You know me: cancer is the thing I fear. If it were just the hot flashes, I could deal with it, as I do not have them that badly, yet. I get hot and sometimes nauseous when I get hot, but I don't get them at night.

My sleeping is not great. I take Klonopin more than I used to, but I would like to cut back on that. The issue is I don't feel like me. I cry soooo easily. I get thrown off my game in a second. It's like I've misplaced the person I was and I can't find her. I feel old as shit, which I so hate. Thank God for Botox. I eat well. I exercise five days a week minimum, so I'm doing all the right stuff. I just want me back. Dr. B [my radiologist] says, "Better HRT than Prozac."

Sex—well, sometimes it sort of hurts. I mean . . . I sort of forget about sex. If Glenn were not here to remind me, I would probably forget about it altogether. I feel like I'm not as sharp, not as funny, not as with it. Sometimes it's like my outline is in the room, but I left my content in the drawer.

I feel like God takes all women at fifty and throws them in a trash compactor. I think that sucks.

This pretty much sums up the way I was feeling. Seeing it on paper and finally owning it outright and then going back to Dr. MacBook and reading some more literature on menopause was pushing me further toward HRT. If you go on menopause websites, you find things like this: "As you approach menopause (or are propelled into it) your most immediate concern is usually the in-your-face symptoms: hot flashes, weight gain, vaginal dryness, mood swings, loss of energy, and skin and hair changes. Long-range issues include osteoporosis, heart and vascular disease, and how to live a reasonably full and happy life."

Hold on—the best you can offer me is a reasonably happy life? I'm not satisfied with the thought of compromising perhaps the next thirty-five years to be "reasonably happy." And I'm certainly not willing to donate the next ten to such a drastic concession.

So I made a deal with myself. (I'm always making deals with myself or the imaginary supernatural being, whoever he/she may be.) This deal was I would go in for my annual mammogram and sonogram, and if everything was okay and my radiologist told me she thought it was okay, I would take the giant step. My pill-taking, patch-wearing, glowy-skinned friends were cheering me on; my militant anti-HRT group looked at me like I'd joined the KKK.

The mammogram was fine. The radiologist assured me she saw no difference in the amount of cancer between women who take it and women who don't, with the exception of those who have a family history of breast cancer. She told me she takes it herself. I repeat, she takes it herself, and she is one of the top radiologists in New York. She spends her days looking at mammograms; she has two kids, a happy marriage, she doesn't want to kill herself. My ob-gyn was also on board, though he made it very clear the decision was mine. He was merely offering me options, and HRT was one of them. My regular doctor, who is an oncologist as well as a GP, warned me against it. My acupuncturist was appalled. My dry cleaner felt . . . okay, I didn't mention it to the dry cleaner; the point is, when it comes to HRT, there are as many opinions as there are people and studies and blogs and personal histories, and they are all conflicting. Eventually, at least in my case, I had to choose what I felt was best for me. My mother was on it for forty years and she was fine; there is no breast cancer in my family; I was so fucking miserable, I could barely get out of bed. I decided to go for it. I went home and took one of the little white pills and threw away the warning pamphlet.

Taking the pill did not alleviate the symptoms right away. My ob-gyn told me it takes at least two weeks to really see a difference. But I watched myself like a kettle coming to a boil. And the truth is that for four weeks a new symptom was added to the mix: raging paranoia and fear. The rational me—or irrational me, depending on your point of view—was taking the pills; the hypochondria-riddled, cancer-phobic me truly felt like each pill was a step closer to my own demise. So now I was paranoid and hot and headachy and miserable. Was there no way out of this?

I kept emailing and most likely torturing my ob-gyn. He kept telling me to wait and it would kick in. I said I could maybe stand the fear if the pills were working, but coupled with my self-inflicted death march, the symptoms were doing me in. The man is very patient and Chinese (not that one has anything to do with the other), but he told me if I was so concerned I should get off them. He also said I was taking such a low dosage that maybe it wasn't enough to be effective. I was so estrogen-depleted I needed a bigger dosage? I had demanded the lowest dosage possible. I think I was taking enough HRT to help out a menopausal mouse. After a full month of little change he upped the dosage. That did nothing but increase my fears until—holy-moly guacamole, as my eight-year-old would say—I started to feel better and better and better. First out the door were the mood swings. Then the hot/cold valve shut down completely. The headaches totally disappeared and my memory and energy came back. I felt positively fabulous. In fact, I felt better than I had in many years. Heck, maybe I'm wrong, stop the presses—fifty *is* the new thirty.

But the medical answer for this was that I had been losing estrogen for years. They say it starts dwindling in your late thirties—that's the reason women over forty tend to have a much harder time getting pregnant. So despite the fact that when you hit menopause it may feel like your estrogen has gone from sixty to zero, it's actually been going from sixty to fifty-eight to fifty-two and on down, until the time when your periods end and your tank is empty. It happens so incrementally that you don't really notice it. But when it comes back, boy, are you on your game. I started taking the low dose in July and in August upped it to the one that worked. I don't know what part the HRT actually played in the story I'm about to tell, as it took months to get a straight answer, but many different events collided at the same time.

In the summers we live out on Long Island, which, if you look at a tick map of America, is tick central, meaning Lyme disease is an enormous problem. In fact, the two most popular topics at parties are real estate prices and who has had Lyme. Lyme is not a joke; if you discover and treat it in time you are fine, but if you don't you can get chronic Lyme, which comes with a multitude of long-range problems. (I will let you consult your own Dr. MacBook if you are interested in the various symptoms, cures, and side effects.) So at the end of August, at the same time as I started taking the higher HRT dosage, I got bitten by a tick and came down with Lyme. I knew I had it, and so, although the doctors like to wait a month to put you on the antibiotic doxycycline to be sure, me being me, I put myself on it right away. (You might be wondering how I pulled this off. I have a huge stash of my own doxycycline, the reason being that I go to India a lot and take it to prevent malaria. I have Cipro, Tamiflu, and five Z-Paks at any given moment.) So luckily for me I had the doxy—you know you are a hypochondriac when you start giving your meds nicknames—and I took it. That took care of the Lyme. I took it a month longer than they prescribed (forget the fact nobody actually prescribed it), but on the hard-core Lyme sites Dr. MacBook says that doctors never prescribe it long enough. So I took it for two months.

Two days after I found the tick I developed strange lumps in my face. These sent me on an around-the-city-in-eighty-days tour of every specialist in New York. The Lyme had triggered some sort of reaction. For months I thought I had everything from lupus to lymphoma, and that's just focusing on diseases that begin with *L*. Then one day while getting a pedicure I noticed a nodule on my leg. The manicurist said, "Everyone has those." Well, I never did. The next day I joined my husband on a business trip to Buenos Aires, where the lump grew and got big and red and started hurting. So I spent much of what should have been an amazing trip staring at it, pushing it, prodding it, and fretting about it. Since I hadn't brought my computer, I couldn't consult with Dr. MacBook. I had a massage every day instead; I thought maybe it was nerves and we could relax it away. I kept asking the masseuse what it was, but she spoke no English. She would just break into tears, which made me very nervous, as I thought that, being Latin and deeply religious, she maybe had a direct line to God and saw something ominous in my future. It wasn't until the last day that I realized I was tipping her what she normally made in a month and she was crying for joy at the end of each session. (I have a hard time with foreign currency.)

One morning on the trip a lump appeared on my arm. It was

smaller than the one on my leg; was all the massaging spreading them out? Was this related to the still-undiagnosed lumps in my face? Was I destined for a life of lumpiness? Was I going to turn into one of those creatures people stare at in the streets?

One night by mistake I missed my HRT; too much Argentinean red, I guess. I slept fine and woke up the next day symptom-free, so I decided to see what life would be like without HRT. Life was fine for three days, and the lump on my leg started to shrink. There was no change in my face, however, and I spent much of my days ping-ponging between my face lumps and my diminishing leg lumps. My husband, who is abnormally patient, finally yelled, "Stop touching yourself!" But eventually I connected the dots. I stayed off the pills the remainder of the trip and the leg lump shrank to the point that by the time I got home it was almost gone—though I was back to being totally menopausal, with a lumpy face. Great. Would I have to stop the pills? Would my choices be lumpy legs and face or suicidal and hot?

I went straight from the airport to my GP, and he asked me if I had things shot into my face. I said of course I did. He said that was what it was and the leg was a bruise. Now, I knew it wasn't a bruise. I told him about the HRT connection, and he told me I was crazy and to go back on the pills. This was the same guy who'd told me not to go on them. Doctors: can't live with them, can't live without them. So back on the HRT I went, and the lumps returned. The story goes on for months; I will not bore you with all the details. But as in any good story, there are heroes—three of them—and a happy ending.

Eventually I decided that since I felt Western medicine was not delivering the answers I wanted, I would turn to the East. So I started doing heavy-duty acupuncture.

My acupuncturist took one look at my leg and brought in the head guy at the Wellness Center. He said, "You have erythema nodosum," and promptly left. (Holistic does not always translate into homey bedside manner.) It sounded beyond creepy, and possibly life-threatening. "What is it?" I shrieked to Ariya, the world's best acupuncturist. She shook her head and went right to Google. Oh my God, real doctors use Dr. MacBook too! She got the answer right away. The description matched my symptoms exactly: red, tender nodules, usually located under the knee and on the chin, though they sometimes could sprout on arms and trunk. Okay. So now that I knew what it was, what causes it? Well, the first thing listed under causes was birth control pills, followed by estrogen. Damn, it was the estrogen. "I told you not to go on those pills," said Ariya. She had been the loudest voice in the stay-away-from-HRT chorus.

So back off the pills I went, and misery, headaches, unpredictable moods, and body temperature fluctuations became the daily routine. I was so pissed off—I had finally found the cure, but I was the one out of tens of thousands who ended up with this obscure symptom. It was so obscure my ob-gyn had never seen a case in thirty years of practice. "Well, it's the second thing listed under causes online!" I yelled at him. I had never yelled at him, not even during childbirth, when I had an excuse.

Now I was triply depressed: depressed because I wasn't on

the pills, depressed because I couldn't take the pills, and depressed because I still had lumps all over my face.

At this point the menopause was back in such full swing I thought my husband would leave me. "I wish you could take those damn pills," he would mutter.

"You and me both," I said, throwing off three quilts as the internal heat turned up.

I continued to go to doctors; I needed answers as to why I had lumps in my face and what if anything could I take for the damn menopause now that estrogen was out. I went to the most thorough blood doctor I could find—at Quest Diagnostics they call him Dracula, because he tests for diseases yet to be discovered. He happens to be our friend and an endocrinologist. He tested me for everything imaginable—those things that could kill and those that were benign. I had nothing wrong with me, or at least nothing the blood work could divulge. But I had a lumpy face and I couldn't take estrogen, and that felt like a lot to me.

The lumps in my face are worth describing, as they led to the ultimate cure. They were on my lower face and stuck out on either side; they not only looked strange but puffed my face way out. Dr. Dracula was stumped, so he sent me to a leading rheumatologist, telling me, "You could have rheumatoid arthritis." Wow—in all my years of worrying about possible diseases and all the time spent on WrongDiagnosis.com, that was one I'd missed.

I told him, "I'm too young for arthritis. That's an old person's disease."

He gave me that look only doctors can give, equal parts parent and deity. "You're fifty."

I thought arthritis was nothing, a little stiffness in the morning, which I had anyway after going off my hormones. As long as I was going to be living on extra-strength Tylenol to combat my daily menopausal headaches, I might as well have arthritis too. Besides, *arthritis* sounds like *sinusitis*, and usually things with an *-itis* at the end don't do you in. But Dr. MacBook told me about rheumatoid arthritis, and it's horrible. Go look it up yourself.

So off to the rheumatologist, who, after patiently listening to the story of the Lyme disease and the Sculptra and the erythema nodosum and putting me through several tests to rule out any form of arthritis, said I was in perfect health, though a little high-strung. Really? The barista at Starbucks could have told you that. I just wanted to know what those lumps were and why I still had some erythema nodosum on my leg, and *I wanted my estrogen back!*

This guy was a saint, the next Jewish saint after Edith Stein and my husband. He said, "I'm giving you prednisone for the nodules on your legs."

"Prednisone? That's the stuff that makes Jerry Lewis look like a float in the Macy's parade."

"Not if you take it for a week," he said. "In fact, it might make you euphoric." It actually did; luckily it has such horrible side effects or I could have become addicted.

Then he said, "The lumps in your face are a result of your

immune system's rejection of your injectables in response to the Lyme—they all fell to the bottom of your face. I will send you to a dermatologist who will take care of it."

"But what about the hormones?" I shrieked.

"Many people react badly to synthetic hormones, but those people tend to respond very well to bioidenticals. I will give you a prescription for those and you should be fine. I will also send you to an ob-gyn who specializes in menopause and she can see you through all of this."

"What if I get the nodules back?" I asked him.

"I don't think you will. If you do, you are one of the very rare people who cannot take the bioidentical hormones, but I don't think that will be the case."

So off I went, prescription for bioidenticals in hand, with no life-threatening disease and the name of a doctor who would reverse my Frankenstein's-monster face. The nodules never reappeared, though I look for them from time to time, as it is not my nature to leave well enough alone. The dermatologist shot up my facial lumps with steroids and told me I was the third case she had seen where the body's immune system rejects injectables when it goes into overdrive, as in the case of Lyme. Which is exactly what the acupuncturist had been saying all along.

And the best part is that the bioidenticals worked—in fact, they worked faster than the synthetic hormones.

Bioidenticals have been around for a long time. They have been the hormone of choice in Europe for decades. But for many reasons, some having to do with large drug companies and the

FDA, they have not been widely used in this country until Oprah admitted she had started taking them. Then overnight, thanks to Oprah's amazing hold on the American female psyche, they became a viable choice for women suffering from the various and unpleasant side effects of menopause.

Suzanne Somers, whose books have sold in the millions and who has been plugging bioidenticals as a cure-all and fountain of youth, has gone so far as to claim they prevent cancer. She has also taken an enormous amount of flak for this. There are many people who feel we should not be taking our medical cues from celebrities. There could be truth to this; however, for me it depends on the cues and it depends on the celebrity. I don't know if I would listen to Ozzy Osbourne if he climbed on a soapbox and started proclaiming the miraculous results he's seen mixing gin and oatmeal, but Suzanne Somers did do a lot of work on herself and is a breast cancer survivor, and Oprah has to be very careful about what she promotes.

What exactly are bioidenticals? According to Dr. Robin Phillips, who edited *The Menopause Bible*, they're hormones that are biologically identical to what the ovaries produce. Furthermore, our ovaries send the most important form of estrogen, estradiol, directly into the bloodstream. But when you take an HRT pill, the compound goes through your liver; this is not the way we get our hormones when we self-produce them. Bioidenticals are absorbed either transdermally (you rub them on your skin or apply a patch) or vaginally, and so they go into the bloodstream directly.

The second way in which they are bioidentical is they are composed of the same chemicals the body produces on its own. This does not mean some guy has to make it especially for you like a latte with extra foam and two shots. There are companies who make bioidenticals that are FDA approved and where the quality control is controlled. So you need to do some research and hook yourself up with a doctor who really knows what they are doing. You don't want just anybody whipping you up a batch of hormones in their blender.

For decades we have been taking our medical cues from doctors and drug companies, who are often in bed with each other. It is a very tightly wound knot, and one the average woman cannot begin to untie on her own. Sometimes we have to turn to friends and media figures who are willing to share their stories and research—again, they are not doctors, but this is one of those choices the individual has to make. Whom are you going to listen to? How does the average woman wade through the river of conflicting information that is out there? It really gets down to two basic questions: how miserable are you, and what are you willing to do about it?

As I write this, I have been on bioidenticals for two years. I rub an estradiol gel on my arm each night. I have no side effects from this, and I feel like myself—and closer to what I was in my early forties. Every four months, I take progesterone pills for two weeks. Those make me feel wonky but are necessary to avoid some nasty reactions from my uterus, which I guess feels ignored; it seems like a small price to pay for feeling like a human being and not an aging shrew. My skin looks better, although it's not perfect—there are no miracles when it comes to aging.

The end of estrogen is permanent. I hear some women get over their symptoms, but I hear more stories from women whose symptoms hightail it back even in their ninth decade. So how long will I take bioidenticals? Truthfully, until someone tells me they will kill me. I can't imagine life without them. (Well, I can, actually, and that is why they will have to be pried from my clutches.) Nothing is a cure-all, and there is no Shangri-la; we are programmed from birth to age. But thanks to science there are things we can take that help us along the way.

Still, the bioidenticals can only do so much. They do not totally revive your libido: they give it a push up the hill, but they do not replace the engine. Chances are you will eventually need a lubricant when you have sex: the lining of your vagina does thin out over time, and it will continue to do so, but again, taking something slows down the process. Also, bioidentical hormones do not entirely help with the sleep problem, but as women get older they tend to stop sleeping as well as they did before. (Studies have proven we don't require as much sleep as we age, though I really like eight hours.) So depending on your individual sleep needs you may require anything from a prescription to just some over-the-counter calcium-magnesium tablets before bed. And lavender sachets on your pillows do work!

Even with the bioidenticals I am not asymptomatic. I get the odd hot flash. I remember my mother, who was on the real deal estro-bomb pills, sticking her head in the freezer from time to time.

I used to think she was searching for some long-lost Häagen-Dazs rum raisin, when in fact she was cooling down from a flash.

Thirty million women in the United States are now in or past menopause, and another six million or so will reach this stage of life in the next decade. People will tell you menopause is one of life's passages and that you should just get through it and move on. I don't know many in the thick of it who would agree. (Sure, it's one of life's passages, but so is death.) Menopause is not only a passage; it is a legitimate physical condition, for many women the first real one-two punch announcing that the aging process has not only begun but is here to stay. It has physical symptoms that can range from tolerable to crippling. There is no question that it ages you and drags you down in many ways. I truly believe if someone came down and asked every woman on the planet, "Do you want to wrinkle up like a prune, have your body run hot and cold like a shower with faulty taps, lose your interest in sex or have it hurt, have your bones get brittle and break easily, watch your hair thin and your memory go, suffer through mood swings that resemble Lindsay Lohan's, and endure headaches, joint aches, and heartaches? Do you want to lose the ability to do something so simple as fall asleep? In plain English, do you want to age virtually overnight?" How many people would say, "Sure, sign me up"? Sounds like a great way to spend the next thirty years.

There are people out there who will say, "Get off it—go with the flow." To them I say, no way. Why shouldn't you treat that condition as you would any other?

Fifty may not be the new thirty, but there are ways to make it closer to forty than sixty. For me that is a much more appealing option than spending the next thirty years drowning in my own sweat.

3

Sex, Estrogen, and Not So Much Rock and Roll

I started out to be a sex fiend, but I couldn't pass the physical.
—ROBERT MITCHUM

When I was thirty I masturbated every day without fail, some days twice. Now that I'm fifty I never do. I didn't actually think this was a big deal. I didn't really think about it at all. In fact, I wasn't even aware that I had stopped. It's not a conscious decision like giving up carbs or switching from jogging to yoga to let your joints have a break. Like many things that drift into the background as you get older, masturbation just sort of fades away without a grand farewell.

It's actually rather odd that I didn't notice, because it was such a big part of my life from my teenage years through my thirties. Throughout my thirties, every day I would drive my daughter to school, come home, lie on the bathroom floor, grab one of my three or four foolproof fantasies, and bippity-boppity-boo, I was done and into the shower, and my day would begin. This was as integral a part of my morning ritual as brushing my teeth or applying mascara.

And then if I woke up in the middle of the night and couldn't get back to sleep, I would have another go. Bippity-boppity-boo—three minutes, happy, and back to sleep.

Now, you'd think the departure of such a routine would be missed, perhaps even mourned. But if it were not for my doctor, the remarkable Ed Liu, I'm not sure that I would have even noticed. One day I was in for a biannual exam (biannual is a good idea after fifty). Anyway, Ed was going over my blood work and studying the plunging estrogen levels when, without lifting his eyes from the report, he said, "Do you still masturbate?" Ed has a no-bullshit character, so this question was totally appropriate.

"Now that you mention it, no—not for some time."

It was clearly the answer he was expecting, though I felt a little embarrassed. It was like admitting you hadn't been invited to the party of the year. God, I really was getting old and out of it. I was the person I had been having sex with the longest, and here I had stopped turning myself on.

"But you and Glenn still have sex?" He wasn't dropping this.

"Oh yeah, totally, and it's fine, I mean, not as much, 'cause of life and the kids and all that. But yeah, me and me, I guess we sort of broke up."

"Well, try to play with yourself from time to time. It's good for you, keeps the juices flowing."

Is that what part of getting older is about—you have to be ordered to masturbate? Seems a tad sad. Pathetic, actually.

Though old people do play with themselves. The first person I ever saw whacking off (or a version of whacking off) was an eighty-year-old man in a rest home. When I was twelve I used to volunteer in an old folks' home several days a week, serving food, reading to people so old they couldn't see, and sometimes just sitting and watching *The Brady Bunch* with ninety-five-year-olds in pigtails and diapers who had been dumped by their families. I could never do it today, now that I'm closer to them than I am to second grade. But I truly enjoyed it and logged about twenty hours a week.

At this home was an ancient yet horny man. He had his own masturbatory ritual: he would throw off his big blue diaper and frantically try to make contact with his truly withered penis. After a few minutes he would find it, then pull on it with all his feeble might, stretching it Slinky-like as far as it would go . . . and then *ping*, he would release it like a rubber band. It obviously wasn't painful; in fact, it clearly was pleasurable, as he could be counted on to do this by the hour, and this guy had real hours to kill. Pull and *ping*, pull and *ping*—all the while with a grin of satisfaction on his face.

I, of course, had never seen anything like it; my guess is that outside of rest homes and asylums, few have. But I would always find an excuse to end up on his floor, inevitably outside his room, where I could watch the show. There is no question that it had a yuck value; it was way up there on the yuck-ometer. But I had seen few penises of any sort at this point in my life, and had never seen anyone playing with anything like this—nor had I seen anyone in the rest home, or perhaps anywhere else, this happy.

I remember the older nurses making some excuse as I stood in the doorway, my face declaring my mixed responses of confusion, fascination, and disgust. "Older people . . . you know, wack jobs, all of them," they'd say, and they would push me and my food cart out of his line of fire.

This was my introduction to masturbation.

And then my beloved grandmother took to a vibrator in her eighties. She didn't just take to it, she became obsessed with it, going through batteries at a rate that would impress the Energizer Bunny. She went through vibrators too—she started out calling them facial massagers, and then at a certain point she stopped making excuses for them and just demanded someone buy her a new one during the next pharmacy run. Of course, whoever went to replenish her supply would ask for a facial massager, as that was the euphemism back then. But the idea that the pharmacist would think Grandma was home diddling herself while watching *The Merv Griffin Show* was too far out of our comfort zone. (Though, quite frankly, we were so impressed that this woman, who had never exhibited even a hint of sexuality, was finally in her very last years finding some pleasure in her body that we just kept her battery supply up and let her massage away.)

So maybe my masturbatory slump was just the fault of the fifties, this period where your estrogen first falls off. But I for one was not going to wait thirty years to get sexually reacquainted with myself. I was going home to masturbate like it was 1989.

The first thing I needed was a new fantasy. The only fantasy I remembered from my thirties was one involving Jack Nicholson wearing only a dinner jacket at the Academy Awards. Jack was no longer a viable option, as he was too fat and old for fantasyland, and since I had given up any hope of ever receiving an Academy Award, I didn't want any bad vibes around. I had to come up with someone new. Whom do I love? Who was hot? Who would get the juices flowing? *Jon Stewart.*

But of course! I am totally hot for Jon Stewart. He's smart, he's funny. I always go for brains and humor over looks and brawn. Not that he isn't cute, but you know, if you gave me a choice between Jon and Orlando Bloom, I would take Jon any day. Perfect casting: Jon Stewart was now the central character in my new and oh-so-fabulous fantasy masturbatory life.

So, just lie back and think about Jon Stewart. This is good—climb back on, just like riding a bike. Jon Stewart, Jon Stewart . . . where would we be? Hotel? No . . . Paris? No . . . We could be in Sag Harbor, we both have houses there—but no, too weird, he has kids, I have kids . . . Where? Shit, I don't really have time to location scout right now. I have to get to work and the cleaning lady will be here soon . . . Jon Stewart . . . Jon Stewart naked . . . he's naked . . . and he's hot and I'm . . . I'm . . . His cock is huge . . . maybe . . . it's throbbing . . . something . . . I'm having masturbation block here . . . I'm something . . . I'm licking him . . . Why would I do that? I hate licking people. Oh, fuck it, I'll lick him . . . I don't know . . . I'm hot and he thinks I'm hot and he grabs my tits . . . Who would grab my tits? They are so saggy. And I really need an eye lift—I have to start saving for that . . . And there is his hot dick . . . Yeah . . . I used to be so much better at this . . . Maybe I need to undress him, like a Ken doll, so I am taking off his pants at this undisclosed location and there is his giant cock? . . . And you know, it's really too bad I can't write on his show. It's in New York, I'm in New York and I'm funny . . . I am really funny, but he would never hire me . . . He only hires thirty-year-old guys from Harvard like everyone else these days . . . Who is going to hire a fifty-year-old woman to write on a comedy show? . . . Big dick, big dick, hot, hot . . . He's fucking me . . . It's why I haven't worked in two years: I'm fifty. I'll never be Nora Ephron; that ship has totally sailed . . . I'll never work again . . . Stay on topic, masturbating, feels good, feels good . . . Jon Stewart naked, with me . . . No work . . . Can't get a job because of guys like him . . . Which reminds me, Lucy's tuition is due next week; have to call Merrill Lynch . . . Jon Stewart hot, naked on top of me . . . Go, Jon, it feels so good . . . . Good to be fifty and out of work, with sagging tits, and I'm not even wet—I used to get wet in two seconds, all I needed was Jack Nicholson without his pants at the Oscars and I came right away . . . Jon at the Oscars . . . I prefer Billy Crystal, not to fuck but he was a much better MC . . . Those songs were hysterical . . . And City Slickers *was a great movie . . . Hot, hot, I am hot—it's hot in here. I'm having a hot flash. What time is it? . . . I don't have time for this, it's so boring.*

Question: if you fake an orgasm and nobody is around, does anybody hear you?

My sexual ADD or lack of follow-through has nothing to do

with Jon Stewart or my amazing husband (who, I may note, is a wonderful, giving lover). And I'm sure Mrs. Stewart—whose name, by the way, is also Tracey—has endless nights of pleasure with Jon, who I'm sure is as dynamic in bed as he is on the air. This is another case where all roads lead to estrogen. And it makes perfect sense—if you are not meant to have babies, why would you be plagued with the constant urge to have sex? The universe is no dummy. It knows it's easier for you to score a handbag than to get knocked up, so the fact that many of your random thoughts revolve around inane things like shopping, tuition, or deciding whether or not to drug-test your fifteen-year-old makes perfect sense.

I also don't lust after young men or other really hot men. As you can see by my feeble attempt to kick-start my libido into overdrive with Jon Stewart, this is more of a case where there is fire in the brain, but the furnace needs a total renovation. You know that feeling you used to get? Well, maybe you're still getting it, in which case you should head to the stadium with the women who love birth pain, got through menopause without hot flashes or mood swings, and never had cramps or PMS—you head on over there now, I think there are plenty of seats. But the rest of you remember that feeling: it was like fire, a thump, a thud, a hot prod that originated deep inside you, starting right above your pubic bone, and then just rising like molten lava up your insides. The female equivalent of a violent hard-on, but felt internally because, after all, we are women. It was that I-have-to-have-sex signal, and if you were in spitting distance of someone who wanted it with you, chances are you had no choice but to give in. At least I did—delayed gratification has never been one of my strong suits.

I personally have not had that feeling in years. And you know what? I miss it. I miss it a lot. It was alive and it reminded me that I was. It was a vital sign—not like a heartbeat or proper kidney function, but it was a real turn-on (for lack of better word) in a myriad of ways.

I had it a lot for my husband when we got married, well up until our daughter was born, and probably several years after. Vacations have a way of igniting it, but I have not experienced this kind of lust in its raw, impulsive state since I went through menopause—and all the concentrating on Jon Stewart or the things that used to work will not bring it back. I sometimes try to magically summon it, à la Samantha in *Bewitched*. (I don't wiggle my nose, though maybe I should try that next.)

I try sometimes on random guys, those guys you see walking down the street that are oh so young and hunky. Perhaps familiarity breeds a type of sexual boredom, and that blue-eyed stranger crossing Third Avenue might wake up the sleeping volcano. *Concentrate on that guy with the piercing eyes, the Hugo Boss tie, and what looks to be a six-pack under his tight shirt. Concentrate: what does he look like naked? Come on, undress him, mentally unzip his fly, keep going, and whooooo—will you look at that purse on the woman in the Burberry coat? I've never seen that shade; it's sort of gray and green while still blue. I want that bag. It would go with everything. I wonder where she got it.*

Now, I exaggerate slightly to prove a point, but the truth is I'm not wandering around lusting after things I don't have at home, with the exception of the bag.

And I want that feeling back—I want to be the lioness I once was and pounce on my truly amazing husband as he walks in the door. But because the estrogen isn't pumping those feelings into my loins, and because as a result of daily life I can't concentrate long enough or hard enough to conjure them up myself, well, it just isn't there like it was.

And I know I'm not the only woman my age who feels this way.

I asked many women over fifty how often they think about sex, and 90 percent said almost never. The *almost* says that sometimes they do; ipso facto, women over fifty do have sex and think about it—just 90 percent less than we used to. And when we have it we like it, much of the time. But, God, it takes so much more work for half the pleasure.

It takes so much longer to get wet, and it takes longer to get a guy over fifty excited, and just when you do, your brain kicks in again: *I really wanted to finish that book tonight; I hope this doesn't take too long.* While we are standing on the honesty stage, we might as well 'fess up: we don't come like we used to either. That same Krakatoa-like eruption, that screaming Ellen Barkin orgasm in *The Big Easy*—it just ain't happening. Maybe with vacation sex you can re-create a relaxed version of what it once was. But on a daily basis it is not occurring at fifty, not in my house and not in the houses of the women I talk to. Now, again, if it's happening to you, if you are over fifty and having the same violent, earthshaking orgasms you had in your twenties, thirties and early forties, then you head over to the stadium too. God bless you, and pop me an email with your secret.

This is just another example of the universe's plan in action. After all, if you were still having those mind-blowing eruptions and looking to constantly throw yourself in bed with your mate or any handsome specimen that walked by, what would be the point? You can't reproduce, and the universe is interested in reproduction and the continuation of the species, not your momentary pleasure. If the universe were interested in your momentary pleasure, raindrops would be made from organic dark chocolate and not recycled dirty water.

I also think the universe's plan has another protection policy at play here. Let's say you were throwing come-fuck-me looks indiscriminately at guys on the street or in restaurants. Chances are the looks would not be returned the way they used to. I've tried throwing sexy looks a guy's way; they tend to reach for their cell phone or turn in the other direction. Most guys don't cruise fifty-year-old women, and if they do, they are in their seventies. So the fact that we don't do it so much saves a lot of hurt feelings. A handbag is within reach, that hot guy crossing the street is not, and trying to get him reminds us we are not as attractive to the opposite sex as we once were, which is not a life-enhancing experience. In fact, it is a real mood crasher. So in a way we should be grateful for the lowered libido. But I think I would take Mt. Saint Helens in my crotch every now and then in exchange for a few hurt feelings.

And then of course there is the other hormone, your man's partner in crime, Tommy Testosterone. And Tommy is saying sex, sex, sex, at any price! Tommy is pumping, pumping, pumping his testosterone juice into men's bodies like oil in the deserts of Kuwait. Tommy is dragging them to Hooters and telling them the girls are really smart. Tommy is sometimes very evil and forces them to leave their wives for a Thai girl young enough to be their daughter. Like many bad boys, Tommy has a good side, but when he's not kept in check he can be a real troublemaker.

Tommy Testosterone is our buddy too, but he doesn't hang around as much since we're not that interested in sports and we don't laugh when he farts. So he tends to spend less time with us throughout our early years and pretty much abandons us altogether when we hit fifty, 'cause old girls make Tommy T. gag. Tommy Testosterone also prefers younger company in his male companions, so he starts winding down his relationship with his male buds when they hit fifty as well. It's a slower departure, as there is a certain codependency at play and he leaves more of his belongings behind, but little indicators of his continual presence, like those ramrod-morning hard-ons, he takes them and hands them out to sixteen-year-olds who don't have to catch the train, drive the kids to school, or deal with mortgage rates. Think about it: morning flagpoles on fifty-year-old guys? You don't see them like you used to. And I'm not just talking about my husband, though he is the only person I wake up with or do anything else with. In my younger days, I slept with fifty- and sixty-year-olds—I can't even begin to explain why, that's another book. Just trust me, Tommy T. abandons them all eventually.

Sex, sexual desire, and the act itself are never easy to lump into one desktop folder. The truth is, there are many reasons why sex in middle age is not as ramped up as when we were younger. And once again, there are examples that completely shoot down my point. I will list them for you so you don't have to sit there and waste your time thinking up all the things I might have left out. There is new-relationship sex; without question this is going to give you more bang for your buck (pardon the pun) than long-time-married sex. There is extramarital sex; that certainly has an extra something you may need to feel like you're back in college, and this is undoubtedly the reason so many people take part in it. Then of course there is hooker sex, which is far costlier than the original price if you get caught. (Paging Eliot Spitzer.)

The price for extramarital sex if you get caught or possess a conscience (paging Tiger Woods) is very high, and my guess is that it's often not worth it. It costs people their careers. It costs people half their net worth—just ask anyone who has gotten caught, gotten dumped, and gotten divorced. That's a very high price to pay for something that can last between two hours and two minutes.

But clearly the lure of some hot time in the sack is worth the risk to many. Though when you try to get the actual numbers of how many actually stray they vary from survey to survey, since it's likely that many people don't 'fess up. Let's face it, it's not like the Nielsens; if you admit to watching *Two and a Half Men*, you're not going to be moving into the Days Inn and turning over your entire 401(k) as a penalty. Thus the precise figures on who partakes and who resists are hard to pinpoint. I've read surveys that claim it's as low as 12 percent for women and as high as 55 percent. I split the difference and go with the study that says 22 percent of women cheat on their husbands.

The one figure that is pretty consistent is twice as many men as women cheat on their spouse. Tommy Testosterone at work again: "Come on, look at the rack on the blonde at the end of the bar. I bet she gives great head. Who's gonna find out?" Tommy T., we often learn the hard way, is not a member of Mensa.

But since spousal sex is the one we're concerned with here, the others really don't count (leaving aside the obvious fact that these extramarital affairs are happening because not enough is happening at home). If we work from the template of married with children and a house and then throw in a few pets and surviving in-laws, there are a multitude of reasons why a session with your mate isn't what it was when you were first together, before your mutual snowball of a life rolled down the proverbial hill and collected all these additions along the way.

One of the biggest sex killers is kids. They might as well come out of the womb with a machete in hand and desex you both while you are in the hospital and can be stitched up immediately. The most common complaint you hear from even young men in their thirties and forties is that once the baby gets there, you can kiss sex good-bye. You can blame this yet again on nature: the need to have the baby makes you horny, and once you have the baby, that need starts dwindling. And if you wait until your mid-thirties to have a baby, your estrogen is starting its gradual descent. But in my experience it comes down to pure exhaustion. Kids are tiring and endlessly needy, and if you add a full day's work on top of it, by the time they are finally asleep you are ready to collapse.

Although at least when they are babies you have the advantage that they can't hear you in the next room—or if they can, they can't identify what it is you are doing. Later, women worry that "the kids will hear us." (It could very well be the new "I have a headache." Since the formula for Tylenol Rapid Release clearly was developed by some lab guy whose wife was riddled by headaches, women needed a new excuse, and "The kids will hear us" seems like a great replacement.)

From toddlers to teens, between their school, their sports, their needs, their mere presence—well, it's just not conducive to being thrown up against the refrigerator and screwing your brains out.

But we are really dealing with sex after fifty, which means that unless your kids are adopted or you had them pretty late, chances are they're teens. And no matter how frisky you may feel at six o'clock while you're having a glass of wine and hashing over the day with your mate, things are different at ten-thirty after you have gone three rounds over curfew and all the clothes on the floor. "Why is your cell phone bill two hundred dollars? How on earth can you text five thousand people in one month? Have you ever made two hundred dollars? Excuse me, what is this condom doing in your bathroom drawer? And why do I have to go in and meet with your teacher if you've got a B average? And will you please turn off Facebook while I'm talking to you. Oh, someone asked you to hold the rolling papers? No, you are not going to Ashley's party—her mother is a drunk and her brother is a known sex offender. Well, I'm sorry I ruined your life; you've ruined my sex life."

After a good two hours of that, does anyone feel like getting laid? Even Tommy Testosterone gives up midway through one of those evenings. I can't tell you how many nights my husband and I started out with the notion that we would close the door early and have some alone time, then, after twelve rounds with our teenage daughter, ended the night by passing out with our clothes on after sharing a Klonopin.

On top of that we have the added stress of work and money—or these days, lack of work and money. And if your parents are alive, they are either driving you nuts or on their way out. Then, of course, we live in the age of meds; most men over fifty take Lipitor, many take blood pressure medication, and antidepressants are gulped down by both sexes like Gatorade after a marathon—all of which takes a giant toll on our already compromised libidos.

I don't know about a lot of you, but I feel really guilty sometimes. I truly love and adore my husband. He is an amazing man, he is so good to me, and he is a giving lover. And sometimes when all I can dish up is potato sex, I feel dreadful. Potato sex is not what you think. Well, I don't actually know what you think, but it's not sex using potatoes, nor is it sex on top of a pile of potatoes instead of a couch. It's the equivalent of being a couch potato during sex, which means pretty much doing nothing but lie there while he does all the work. It's when you don't really have the energy to put in any effort but you figure for his sake something is better than nothing. Potato sex is like a ponytail on a bad hair day; it does the job, but not well.

There are couples who have the role-playing game down, and apparently that works for them. But the idea of pretending we're Sonny and Cher and singing "I've Got You Babe" while dressed in a halter top and some crotchless bell-bottoms is just not remotely appealing or sexy to me.

I have girlfriends who have taken pole-dancing classes to spice up their sex lives. One slid down the pole, hit her chin, and ended up with stitches. It's not an easy dance to master, and I don't think it's meant for those over twenty-five. (Frankly, the prospect of seeing certain women shimmying up or down a pole is quite frightening.) And then, of course, let's say you are one of the few menopausal women who master the fine art of pole dancing. What do you do next? You have a pole installed in your bedroom? Not only is it unsightly, but how do you explain it to the kids? Mommy and Daddy decided they want to hoist a flag in their room? Or do you rent out a pole-dancing parlor for the night? Are there special hotel rooms you can book that come with poles instead of showers? It seems to me to be one of those skills to nowhere, like having a degree in eleventh-century Celtic poetry—how are you actually going to put it to use?

Then there are all the belly dancing fools, the women who learn to jiggle and roll their fat in an effort to keep the flame alive. I know very few men who find a jiggling fat stomach erotic. The only ones who truly like it are Bedouins, and let's face it, they see something in camels the rest of the world misses.

Despite the fact that I may not agree with the method, I admire the madness and the effort. You get a big A for effort no matter how you try to pull it off.

Last year around Valentine's Day I was feeling especially guilty, and not only was I feeling like Glenn deserved better, I felt I did too. If it takes two to tango, then it certainly takes two to screw like bunnies. And despite the fact we are in our post-bunny years, I could at least make an attempt at bringing some of that excitement back into our bedroom. So I went about it in the way I know best: I shopped. I decided I would stock up on gizmos and gadgets and toys that would raise our fun and friction to new heights.

There is a very posh sex store in New York. It's not your creepy uncle's sex store in the bad part of town, where you feel like you need a shower right after you leave. It's more like Dolce & Gabbana; it makes sex seem chic and stylish, and if you don't partake and play it their way, you are so last season. (They didn't realize I was there so that I wouldn't feel so last decade.)

The salesgirls are dressed like really high-class sluts, the type all men dream about: not so slutty you can't bring her to the office Christmas party but slutty enough you know she will drive you wild. They have names like Havana and Mignon. And they send out a vibe that says they will *never* offer up potato sex. They teeter about in those shoes that are a little bit Manolo Blahnik and a little bit S and M. And they sell everything from DVDs of upper-end porn like *Last Tango in Paris* to really sexy, elegant lingerie and the latest, most stylish sex toys you've ever seen. They're like sex toys that could have been used in *Blade Runner* or really high-end brothels in Paris or Tokyo: sleek and shiny and, much like a cell phone, capable of all sorts of functions most of us are totally unable to access. These are not my grandmother's vibrator—excuse me, facial massager.

I went home with a shiny black bag that looked like it came from any boutique on Madison Avenue, only it was filled with things that I had been promised would change forever the way I had sex. That is some promise; though this stuff is not cheap, it's a lot cheaper than a weekend in Bermuda.

That night I bribed my younger one to go to sleep early; my older one is so wrapped up in her own world, we could burn the house down and chances are she wouldn't notice unless her computer caught fire.

I put on something black and slinky and lined the toys up on our pillows. Glenn's look told me I had already gotten an A for effort.

I put on a goofy, fake sexy voice that, while it made me sound like an idiot, seemed to go along with the playthings. "Have I got some fun in store for you," I purred seductively, picking up the "Ferrari of vibrators." It was long, sleek, and shiny, with a tiny round control panel that had worked like magic when Havana adeptly slid her finger across it, lightly pressing different spots that signaled it to go faster, slower, jiggly-wiggly, rolling waves—this thing had a colossal playlist. But I couldn't even get it to turn on. I poked and pushed and prodded. Nothing.

"You can't believe what it did in the store."

"I'd like to see." God bless him; he was keeping the mood going.

"The girl in the shop got it to work perfectly."

"Maybe you should have brought her home with it." My guess is he was half kidding. Had he seen her, he would have been dead serious. "Are there directions?" Men are a practical lot.

As with all directions, I'd thrown them out because for some reason, and despite the fact that I can work nothing, I somehow feel I am above manuals of any sort.

"I'll get them out of the trash." I trundled downstairs to rummage through the garbage. So far this was not going according to plan. And what was I doing buying the Ferrari of vibrators when I still didn't know how to drive a stick shift?

Directions retrieved, I was back to bed and ready to roll— though in keeping with the hipness and sleekness of the vibrator, the directions came in a tiny matchbook-size pamphlet that was clearly meant for people who don't need glasses to read.

"Dammit, where are my glasses?" Now, if there is one question I ask more than any other except "Are you sure I can return it?" it's "Where are my glasses?" I have three pairs; they are bright orange and pink. If they are not on top of my head (where they spend much of their time) or on my face, I don't know where they are.

Back up to search for glasses. "Your tote bag is downstairs," Glenn said. He started fondling the motionless vibrator as I schlepped back downstairs. I should have thought of all this before, I guess, but the checklist for a sexy night was vibrator, hot outfit, and pulsating cock ring. Call me crazy, but I left out instruction manual and reading glasses. At this point, I was feeling like I should have added defibrillator to the list. When you're young, all you need is two bodies and a floor, but now it felt like I was going through the checklist for a rock climbing expedition.

"They're not here." Now the obligatory twenty-minute search for the glasses, which were finally retrieved in the wash, folded inside my T-shirt, where they must have gotten caught when I pulled it off my head.

Then back upstairs, reading aids in hand. How not sexy is a fifty-year-old woman attempting to be sexy while wearing glasses and trying to read a vibrator manual? Even with my glasses I was having a hard time with the tiny print; I held the booklet right in front of my face and squinted while barking instructions to Glenn. "If you gently rotate the triangle on the left side of the motion wand it will pulsate in a rolling yet frisky manner." He looked at me like I was nuts, but clearly he was so sick of potato sex he would try anything.

He hit the button. "They say gently rotate, not pummel!" Playing General Patton was not part of my seduction plan, but

I was getting frustrated. He tried a little more gently, but still, nothing happened.

"Okay, try . . . God, I can't make out this word; you need a magnifying glass to read this thing."

We pressed and prodded and rotated and caressed that stupid little button for so long we probably could have flown to Chicago. At one point there was a tiny bit of movement, just a flicker on and off like a lightbulb that is on the verge of expiring, but we were so thrilled we had gotten the Ferrari to turn over that we both yelled, "It moved!" You would have thought our child had taken its first step. He grabbed it, but by the time it got to me it was done. Motionless.

"Even Ferrari turns out lemons." One of Glenn's many great strengths is that he is always looking on the bright side and doesn't blame. Me, I was blaming myself for blowing a hundred bucks on the stupid thing. "You said you had some other toys?"

"I do." I was so busy planning how I was going to chew out Havana and demand a full refund for this faulty piece of crap, I had forgotten the big enchilada of my purchases.

Glenn tossed the disappointing vibrator to the side, and I brought out a little box, which, according to Havana, was the ticket, their most popular item. I don't remember what it was called, and when I went on the website to find it, it had clearly been discontinued, as it no longer exists. The easiest way to describe it is to say it was a rubber-band-type-thing ring that went around the penis and on either side it had a tiny metal barrel type that when turned on would vibrate against the man's testicles and the woman's clitoris simultaneously. You may laugh, but someone spent time thinking this up and it makes perfect sense, both people benefiting at the same time.

So luckily, since I hadn't removed it from the box, the directions were there. This one required assembly. The pieces were so minuscule, they were hard to see even with my glasses, so Glenn held a flashlight while I tried to pry and push the tiny barrels into their even tinier slots on the hard-to-hold-on-to rubber band.

This was so not foreplay; it reminded me much more of torturous hours spent assembling Barbie's Dream House on Christmas morning.

With a little patience, Glenn's flashlight guiding the way, and actually following the instructions, within a half hour we had it up and running. When we pushed the button it worked. Okay— fun, fun, fun, here we come. So onto Glenn it goes, and despite the fact that we were now an hour and a half into this project, we were frisky, and I think—I *know*—he was thrilled our potato days were over. I have to admit the thing worked: it was sort of pulsating me and it was pulsating him. And like the two noisy people we are, we were discussing its merits and how much fun we were having, and you could hear the little hum in the background, and then all of a sudden I didn't feel anything.

"Mine stopped," I said.

"Mine's still going," he moaned.

"Okay, but I want mine to work too. This is a group activity."

So we stopped and felt down—my little vibrating barrel was

gone. It had slipped out of its little sling. Back on went the lights and the glasses as we searched through the sheets, me muttering the whole time about how much it cost. At this point I was sorry I hadn't gone for the pole dancing. There was no little barrel to be found: not in the bed, not behind the mattress, not inside the pillowcases, though how it would have flown off and ended up inside a pillow case I have no idea. But when you are on a serious search mission you look in the damnedest places.

We were on all fours—well, all eights, really, as there were two of us, Glenn still with his vibrating barrel attached (at least one of us was having a version of fun). Nowhere on the floor. "It couldn't just disappear." I took the flashlight—who knew that would become the savior of the evening?—and aimed it under the bed, and there, nestled behind the Container Store sweater holders, was our Chihuahua, Lola Falana, pulsating barrel gripped between her tiny paws. She gave me a look that said, *Just try to touch this, lady,* punctuated by a snarl.

"Lola has it!" I yelled. "Go cover the other side of the bed or she will run out with it."

"What?" He was already back in bed.

"We have to cover both sides."

"For God's sake, just grab it."

"She'll bite me."

"Tracey, she weighs five pounds. Just get it. I have the gym at seven."

I was on the floor, halfway under the bed, wearing nothing but reading glasses, wielding a flashlight in a feeble attempt to convince a possessive, snarling Chihuahua not to swallow the pulsating clitoral barrel. My husband had gone from adoring and grateful to bored and disappointed. The night was *so* not going according to plan.

I lunged for it; she tried to bite me, then picked up the little barrel in her mouth.

"She's going to swallow it," I yelled.

"What do you want me to do?" He was moving into the land of almost pissed off, a place he rarely goes.

"I want you on the other side, covering." I ordered. You'd think we were defusing a bomb in Kabul.

"What's the worst-case scenario?" he said.

"She swallows it and not only do we have to take her to the twenty-four-hour medical clinic but we have to tell them she ingested a clitoral stimulation barrel. She will then have emergency surgery, like when Ramu [our French bulldog] ate the Lego, and it will cost a thousand dollars."

He was down on all fours guarding his side of the bed before you could say, "Do you take insurance?"

So there we were on what was supposed to be our maiden voyage back to the land of hot, wild sex: stark-naked, flashlight in hand, crouched on either side of a bed, guarding a five-pound dog who held all the cards.

I talked sweetly to her. "Come on, Lola, Mommy loves you."

I talked sensibly to her. "Come on, Lola, you don't want to eat that filthy metal thing; it will make you sick."

I tried to bribe her. "Lola, if you give Mommy her toy back, she will go out and buy you that rubber shoe that said 'Jimmy Chew' you liked in the pet store."

I yelled at her. "If you don't give me the damn thing, I'm going to wring your bony neck."

She wouldn't budge.

I even went so far as to sing her song. Despite the fact that it drives the rest of my family bonkers, it makes her very happy— or, like other loony dog owners, I have convinced myself she likes it. I sing it to the tune of Barry Manilow's "Copacabana."

"Her name is Lola, Lola Falana, the cutest dog from Tijuana. Her name is Lola, Lola Falana, dog food and laughter are all that she's after 'cause she's Lola Falana!"

Nada. She just growled and put her paw over what was now her barrel.

"The song didn't work," I said despondently.

"It's a stupid song. We've all been telling you that for years. She probably hates it."

"It's not a stupid song. Jack wrote it." (Our friend Jack Feldman.)

"Even Jack would tell you it's a stupid song."

That's how far the evening had devolved: we were now held hostage by a Chihuahua, debating the merits of a thirty-year-old Barry Manilow song.

But, like in so many other parts of life, when you stop pushing something, it often releases on its own. While I had moved on to the subtext of "Mandy" and was trying to convince Glenn how it had really spoken to me in a very profound way when I was sixteen, the backup troops had arrived in the form of our two other dogs, Sophie Horowitz and Ramu Gupta. They had been sleeping downstairs, but when they woke up and realized Lola was getting our undivided attention, there was no question that they were going to put a stop to it. They stormed her, and as Lola fled from her covert position, I went shimmying under the bed to make sure she hadn't taken the prize with her. There it was, next to some old Starburst wrappers and an earring I had lost months before. It was glistening with dog slobber and covered in tiny teeth marks, but blessedly not inside her.

Sometimes it takes a dog to win a dogfight.

There is no doubt the whole toys-run-amuck thing took the oomph out of the evening. And the problem with going way out on a limb you may not belong on is that when you fall off, you tend to feel a tad foolish. The feelings run from "better luck next time" to "maybe we should simply put a little more effort into what we normally do" to just plain "maybe we are too old for the hijinks of yesteryear." Though deep down I don't buy that. And the truth is, we didn't need hijinks in years gone by. We came with built-in hijinks: a flat surface, some passion, and our raging hormones were all it took. A few battery-operated toys don't take the place of those hormones or the raw passion of youth.

You cannot avoid the fact your body is slowing down in a variety of ways. It's more confusing now; in this new era of the cougar, a fifty-plus woman can look pretty darn hot. Often she looks like she is having the same sex she was at thirty. And depending on whom she is having sex with, perhaps she is. But now that we look better, I think it's almost expected of us. At fifty nobody expected my grandmother to be having sex, wanting sex, or, God forbid, strapping on a dildo in an attempt to bring it all back. At fifty my grandparents had separate bedrooms. At fifty my grandmother's fantasy life no doubt centered around that coconut cake.

Yet at fifty, my mother, who was single, was having affairs. I never asked her about the quality of them. Though she probably would have told me.

The truth is that though my mother looked a lot better than her mother at fifty and I look a lot better than my mother at fifty, our bodies are essentially doing the same exact thing. Taking hormones can help only so much. As a friend in the gym said the other day, "After menopause, none of it is ever the same." This swims against the tide of "it's all fabulous and the problems are in your head," but it's true. It's not all fabulous and it's not in your head; it's in your vagina (or in some case not in your vagina), and it's because the estrogen is not pulsing through your veins. And that is something you can't avoid no matter how hard you try to cook the books or how many trips to the toy store you might make.

Despite all that, I think it's important to give it the old college try from time to time. Just don't forget your glasses, a flashlight, and perhaps a first-aid kit, in case the dog actually bites.

4

If I'm Thirty, Why Do I Need a Colonoscopy?

If you get really sick before fifty, it's either bad luck or bad genes. After fifty—shit happens.
—Dr. Edward Liu, MD

When I was thirty I had four doctors in my address book: a gynecologist, a dentist, a GP, and a vet. At fifty-two I have thirty-five. I have the above-mentioned group, though I now have two gynecologists, one on each coast (I work in LA a lot); I also have a cardiologist, an endocrinologist, a gastroenterologist, three dermatologists (one regular dermatologist for vanity-related issues, one who specializes in lasers for a skin disease my husband came down with at fifty, and one to do the annual mole check). I have a dentist for us and one for the kids, an orthodontist, and an oral surgeon for the odd root canal. I have a radiologist for mammograms, a rheumatologist, a plastic surgeon, a pediatrician for my youngest and an adolescent specialist for my oldest, a pediatric gastroenterologist—oh, and a pulmonary specialist for my older daughter's and my asthma. There are two ear, nose, and throat doctors, one who specializes in acid reflux and one for the basic blocked sinuses. There are two orthopedists, one for the kids and one for us. I have a psychopharmacologist and an opthalmologist, plus in the last nine months we have added a electrophysiologist and a radiologist who specializes in cardiac imaging. My French bulldog has an orthopedist as well as a neurologist. (In dog years he is not yet fifty, so we chalk it up to bad genes.)

My husband informed me the other day that the endocrinologist says we need an urologist. I said, "Why?" He said, "Over fifty, everyone needs one." So we will add him to the list as well. We also have a couple of shrinks, a nutritionist, and an acupuncturist. I don't want to make us sound totally loony or in dire medical shape; we don't use all these people on a regular basis, but they have all entered our lives and my address book at one time or another.

I already had to kick several people off the *E* page of my address book to make room for all the doctors spilling off the *D* page. The odd thing is that even though I have a lot of "ologists" in my life, I feel pretty good—or perhaps I feel good because of them all. Do I feel the same as I did at thirty? In some ways better; actually; if you remove the menopause, I feel almost the same. I do suffer from acid reflux, which used to just be indigestion. Now, of course, something historically thought of as benign if left untreated can cause hideous damage later on. And quite honestly, I'm not sure what fifty is supposed to feel like.

I guess the question is, if we are feeling well, why do we have so many doctors? It seems like every doctor you go to feels he or she needs to send you to someone who knows more about a particular thing. Didn't Marcus Welby take care of everything? And if he couldn't, James Brolin cruised in on his motorbike and picked up the slack.

What I do know is that come fifty, we are supposed to get tested on a regular basis for a multitude of possible health disasters. This indicates we have arrived at the place where—well, shit happens. My friends run the gamut, from amazing health to so-so health, chronic problems, cancer in the past, cancer in the present, and, on the far end, dead. Every day you hear of someone new who has been diagnosed with something dreadful. You say a prayer for them and one for yourself and your loved ones, and then you live your life. What else can you really do?

For whatever reason, fifty is this magical number when it comes to health. The age range between fifty and sixty-five is sometimes referred to as to as the Bermuda Triangle of health. It's pretty evident what that means: it's a big dark hole that out of nowhere just sucks you into its vortex. Apparently if you avoid the triangle altogether, you can trundle on into your eighties. While part of me buys into the theory that these fifteen years are without question the age where good health can no longer be taken for granted, if you make it to sixty-five okay, the getting to eighty does not always follow; I know plenty of people who succumb in their seventies. I suppose it's just a flashing sign that says Danger Ahead. In the animal world it's referred to as culling: the less likely to survive of the species are taken out so the strong can survive and make the herd that much tougher. There is a Darwinian aspect to it all as well.

For those who are trying to keep a positive attitude and even pretend they are actually getting younger, it's not easy given the daily reminders we get from those who are actually sick, the people who know someone who has been diagnosed with something deadly, those who know someone who knows someone who's sick, and on and on.

Once I told one of my many doctors a story about a friend of mine who was dying of cancer at the age of forty-six. His response was, "We all have friends and some of our friends aren't doing so great. That's the way it goes." I guess you cannot be a doctor if you don't accept that shit happens. That doctor is no longer on my list.

And if you are fifty, there is no escaping it. My daughter's fourth-grade health form asked, "Has anyone in the family had cancer, heart disease, or a serious illness?"—not sure what category they put the other two in—"or sudden death before fifty?" What they are not so subtly saying is that if you come down with any of these things before fifty it's abnormal, but after fifty it's to be expected. And then of course every other commercial on TV is about someone with some ailment, and because the actors they have hired to play them are all in their late forties or early fifties, they look just like you. From prime time sports to morning talk shows, pharmaceutical ads seem to be outnumbering all others. One day on *Good Morning America* I counted four in the one hour I watched—one distressing reminder every fifteen minutes of what your future could hold. Not a great way to start the day.

I watched TV incessantly throughout my childhood and adolescence. I have been watching *Today* and *GMA* my entire life, and I can't remember a time when there were ads for so many prescription medications, especially those with dastardly side effects. I remember "Charlie says, 'Love my Good & Plenty,'" not "All around the world, men with ED have taken thirty-six-hour Cialis."

Then there was "Two times the flavor, two times the fun, Doublemint, Doublemint, Doublemint gum." A hell of lot catchier and more uplifting than "Another heart attack could be lurking."

How about "Does she or doesn't she?" for Clairol hair color, as opposed to "Depression hurts" for Cymbalta?

And what would you rather see, a big funny-looking green guy bellowing "From the valley of the jolly (yo, ho, ho) Green Giant," or a fifty-seven-year-old heart attack survivor earnestly confessing, "My doctor told me I should be doing more for my high cholesterol. What was I thinking? Now I trust my heart to Lipitor."

And my all-time favorite: "Say hello, say hello to Poppin' Fresh dough," a happy, cheery commercial that made you want to immediately go out and buy a roll of ready-to-bake biscuits instead of running out and getting a full body scan. I'm sure today they would have the Pillsbury Doughboy on Plavix, for heart disease, and they would tell you that the side effects are severe allergic reactions (rash, hives, itching, difficulty breathing, tightness in the chest, and swelling of the mouth, face, lips, or tongue), bleeding in the eye, change in vision, change in the amount of urine, chest pain, dark or bloody urine, black tarry stools, unusual or severe bleeding (e.g., excessive bleeding from cuts, unexplained increased vaginal bleeding, unusual bleeding from the gums when brushing), loss of appetite, pale skin, seizures, severe or persistent headache, sore throat or fever, speech problems, unusual bruising, weakness, unexplained weight loss, and yellowing of the skin or eyes.

Personally I prefer "A perfect crescent roll every time."

Now, I know the crescent roll could give you diabetes, and the fat content is high and that could raise your cholesterol, but talk about not growing old gracefully—this is growing old more fearfully than we have to. I'm all for prevention and intervention, but do I have to be reminded of it 24/7?

The only medical commercial I recall from my childhood is "Excedrin Headache 110." There is a reason for this: drug companies were not allowed to advertise prescription medication on TV until 1995, which was oh so conveniently around the time the oldest boomers were forty-nine—a year away from fifty and the onset of chronic diseases. America's largest generation and biggest spenders were likely to be buying more meds than Niblets corn and candy.

The fact that the perpetually peppy Sally Field, who will always be the Flying Nun and Gidget to me, is suffering from bone loss and taking Boniva to counteract it makes me so despondent I could—well, use a good dose of Abilify, one of the other widely advertised drugs that combats depression. But this is where they are so clever: despite her age, who is more eternally young than Sally Field? If it could happen to her, it could certainly happen to us, so we'd better get that bone density test and start the Boniva before you can say "double hip replacement" instead of "Doublemint gum."

But we should look on the bright side. Just because Sally's bones are crumbling doesn't mean everyone's will; maybe hers took a real beating with all that flying and getting knocked around by the surfboards. Maybe Moondoggie had a thirty-six-hour erection and banged the hell out of her and that caused her bones to become brittle. But actually that couldn't have happened since they didn't have Cialis or Viagra back then, and Moondoggie was young, so erectile dysfunction was probably not an issue (though premature ejaculation might have been, but they won't bother bombarding us with ads for that because it's not a big problem for those over fifty).

I make jokes because that is my way, but the truth is we find ourselves again at yet another odd place: on the one hand, we are told fifty is the new thirty, but if you watch the relentless ads on TV we are told we either are depressed, suffer from heart disease, have bone loss, have to deal with erectile dysfunction, or must cope with a leaky and overactive bladder, and if we don't now, we will soon. The people in those spots don't look a day over fifty-two, but they are darting to the bathroom so they don't wet themselves. Pass the Paxil, another widely advertised anti-depressant; I may be needing those Depends soon.

The really scary thing is the drug companies are now spending twice as much on advertising as they are on research. Considering that one out of every five Americans will be age fifty to sixty-five—smack dab in the waters of the Bermuda Triangle—by 2015, I think their money would be better spent on helping us rather than hustling us. The medical stats for this set are not great—seven out of every ten Americans in this age group have at least one if not several chronic health conditions. Those conditions range from heart disease and diabetes to chronic back and stomach problems. So if you are fifty, chances are something is bothering you some of the time.

If you happen to be one of the lucky ones who aren't plagued by multiple illnesses, you end up spending an inordinate amount of time worrying about them—although you may not be worrying about the right things. Though all the data tell us women have a far greater chance of dying of a heart attack, almost every woman I know is convinced that it is breast cancer that will ultimately do her in. I don't know any women who are remotely concerned about heart attacks, or strokes for that matter. And while heart attacks kill many more of us than any cancer including that of the breast, and we are far more likely to get whacked by a stroke than our male counterparts, we seldom ponder these ailments as part of our future. We never worry about broken hips, which are far riskier and more prevalent for women than ovarian, uterine, and breast cancer combined. Thirty-three percent of boomer women will break a hip. And seven out of every one hundred of those women will die within a year of that injury. That's a big number—it's higher than the percentage of women who will die from breast cancer. But we don't think or talk much about those ailments, much less walk for them or buy lipsticks that provide proceeds to them; we pretty much focus all our medical paranoia and attention on breast cancer.

In our defense, you cannot find a woman at fifty who doesn't have at least one relative or friend who has had breast cancer. Many of us have lost people we love to it. Right now, without digging very deeply, I can come up with twenty-five women I know personally who have either survived it, are battling it, or have succumbed to it. And if I play the one-step-removed game I could probably reach sixty. I have yet to lose any woman I know to a heart attack, much less a hip fracture; that will come in time, but fifty is usually still too young for that. I know of no other disease that has afflicted as many women my age as breast cancer, and I'm sure that that means every other fifty-year-old woman can say the same thing.

While it is true more women will be diagnosed with it than with any other cancer, most women will not. Twelve percent will suffer breast cancer in their lifetimes. That is far from everyone. Yet it doesn't feel that way. It feels like everyone has it or gets it and that we could very well be next in line.

Ask any basically healthy woman the worst day in the year, and (excluding the anniversary of a loved one's death) nine times out of ten she will say, "Mammogram day." I start worrying about it a good three months before the actual appointment.

Sometimes I play a game and shift the appointment around—if I have a big event I want to be in a good mood for, I will put it off, thinking, *God forbid they should find something, I don't want it to ruin my week.* I never have them around the holidays, as I have always felt that would be the ultimate bummer—you get the news in November and destroy everyone's Christmas. Spring doesn't work; I used to get them in the spring, but my birthday and the birthday of one of my daughters are in May, and I would hate to put a damper on those occasions. And then the fall—well, there's way too much going on then, what with back-to-school, my husband's birthday, and usually the beginning of a new project. I kept moving my yearly appointment around until I got the right time, June or July. We don't tend to travel that time of year, so if something goes awry I have the summer to deal with it.

So it's June or July and I make appointments only for the eighth, or the sixteenth if they don't have the eighth. I do this because eight is my lucky number. If the eighth is a weekend, I book the sixteenth, eight plus eight—double eight means double your chances, double your luck, good mammogram day? I hope.

All women of a certain age understand the mammogram terror. There is a look we give each other; it's our secret-society-of-fear look. We whisper under our breath, "Tomorrow is mammogram day."

Then the person you've told gives you that sympathetic look

and says, "Oh shit, really? I had mine a month ago. Thank God it was okay. You want me to go with you?"

There is no other routine doctor's appointment or test that people volunteer to accompany you on as often as a mammogram.

"Tomorrow I'm getting my moles checked."

"Really? Want me to come?"

Never happens.

And no one offers to go with you to get a colonoscopy.

But because we all suffer from this collective fear of breast cancer, it's just a given that a good friend or even a not-so-good friend will go with you if you want. And a not-so-good friend instantly turns into a great friend if she makes the offer.

I never want anyone to go with me. My husband offers, and I shoot him down immediately. "If you go, then it looks like I have cancer and we're worried." If I ever see a husband in the waiting room of my radiologist, I know the woman has gone through it and is in there for a follow-up appointment or is about to get bad news.

If you take a friend, it's not as bad as taking a spouse, but it still looks like you're really afraid or you expect to need her. I don't want to look like that, nor do I expect to need anybody, though deep down I always fear I will.

For me it's important to march in there despite the fact I'm a nervous wreck and pretend to be strong and fearless. If I act as if nothing is wrong, I figure, nothing will be.

And the entire way there I am making deals with God. "If you make this mammogram clean, I promise to devote more time to charity work. I will not get impatient with stupid people. I will do anything you want. Just give me the word—no lumps in my breasts and I'm at your service." So you galvanize your internal army of strength, you have a heart-to-heart with whatever force you believe in, and then you walk into the office, and all the women there waiting their turn look like dead women sitting, even though three-quarters of them haven't even been seen yet. Most of the time they are all fine, but harboring the same fear: this time could be it.

People in this situation don't really smile or make small talk; they are too immersed in their individual panic. They pretend to read magazines. I assume they are pretending because that's what I'm doing. I pick up *US Weekly* and turn the pages, but while my eyes process the photos, the fearful chatter in my brain is so loud I'm sure everyone in the room can hear it: *Look at that picture—Gwen Stefani's kid is wearing goofy pants. Why would she do that to him? . . . If I do have it, maybe I will have the type that only requires a lumpectomy . . . Bronx—what kind of name is that? Celebrities should not be allowed to name their children: Bronx Wentz, Brooklyn Beckham. What's next, Staten Island Gaga? . . . So you lose your hair. I have a friend who looked better with her wig than with her real hair. You can get wigs that are amazing . . . I wonder if Angelina really pays attention to those kids or it's all a big photo op . . . People go back to work with it. Look at Robin Roberts—she was on GMA the whole time she was getting chemo. And she looks great now . . . That Heidi Montag—one more round of plastic surgery and she will look like Joan Rivers . . . They say 80 percent of all the lumps they find are benign, so even if they find one, that is still a four-out-of-five chance that it will be nothing . . . George Clooney with another waitress. What is it about men and waitresses? . . . Even if it is something, almost everyone I know has lived—it's a manageable disease now . . . Sometimes it is—not all the time. A lot of women die . . . I could die. Then I would never see Lucy and Taylor's children . . . What if they name one of them Bronx? If I were alive, I could stop that . . . If I live, I won't care what they name him. I'll just call him Onyx . . . No one in my family has breast cancer. Family history is very important . . . I think Christina Applegate looks better in her dress than Bethenny Frankel does in hers . . . Wait, Christina Applegate had breast cancer, and look at her and Melissa Etheridge. They're both doing fine. And Kate Jackson and Jaclyn Smith. It's kind of weird when you think all three Charlie's Angels got cancer and the ones who are alive were the ones with breast cancer. That's a good sign . . . I wonder if their set was near high-tension power lines or something . . . Farrah died of cancer . . . I mean, three people in one show got cancer and the show only had three people in it. You never saw Charlie—I wonder if he is alive . . . I thought one in three got cancer. But that's three in three . . .*

This is the kind of crazy journey my mind takes while I'm sitting there waiting for my mammogram.

Blessedly, just about the time my head is about to explode, they call my name and I jump ten feet because despite the fact that all indicators point to the fact I know exactly where I am, the self-inflicted torture trip I'm on has sent me to this other world where reality, time, and space are totally abandoned and I am completely submerged in my most primal fears.

The waiting room is bad enough, but then you go into the antechamber, the small room where you take off your clothes (gown open to the front) and wait for your turn. At this point I'm way too nervous to even pretend to read. For some reason the women who actually take the images all have the same demeanor. I guess they are trained—look blank, say little, stay neutral, do not get emotionally involved with the patients. This makes it virtually impossible to get any information out of them. In the last fifteen years of getting this test (since I was thirty-five; Ed Liu always said it was better to be safe than sorry) I have tried every way I know to get them to respond, and nothing works. They are like the guards in front of Buckingham Palace, except they are middle-aged women, often from Eastern European countries, who spend their days squishing other women's breasts and witnessing their panic.

While they are mashing my breasts between the plates (which, weirdly, has never bothered me as much as it bothers most people), I'm asking questions.

"Everything look okay?"

"I haven't started yet."

"See anything suspicious?"

"We have to take a picture before we can see anything."

Not very comforting, but it has a certain logic. Then she takes the first picture.

"Now do you see anything wrong?"

"I only take the test. The doctor reads them."

"Yeah, but you see hundreds of these a week, which translates into thousands a year."

"Move an inch to the left."

"But if you saw something out of the ordinary, you would know what that looked like."

"We are going to do the right side now."

"Like I know a bad script when I read one—the first two lines and I can tell you if it sucks or not—so my guess is you can just take one look and see if I have a lump, or a bump, or an inverted pimple, or a benign growth."

"Go wait in the small room. We will let you know if we need more shots."

And back you go to the antechamber. I always leave the door open a smidge hoping to overhear them talking to each other, but it's usually about *American Idol* or something totally banal that has nothing to do with my mortality.

I think that second sit-down in the antechamber is the worst. Now, I go to someone who reads all the images herself and does it while you are there; at other places, you go home and wait for the phone to ring. I would rather devote the entire morning and know what's up. But in my case that second wait is when they are figuring out if you do have something wrong and they need more tests to confirm it. It's the guilty-until-proven inno-cent moment, and the part where I think my heart is going to beat right out of my chest.

Every now and then they come back and say, "The doctor would like some more images." I would like some more images too, of me somewhere else, like in Starbucks with a latte, but back I go to the photo room, my feet dragging.

"You found something wrong?"

"The doctor wants to clarify."

"Clarify what?"

All you get back is the Beefeater stare.

Blessedly, that part is often just because I moved or the film registered a shadow. At the office I go to, the best antechamber outcome is, "You can put your clothes on." When they say that, I pretty much know I am out of the woods.

There is nothing like the feeling when your report comes back clean. You bounce out of the office. It may sound overdramatic, and of course the elation is always qualified—it's a temporary victory, a reprieve for another year, a momentary stay of execution—but I think that is how most of us feel.

There are four stages of our relationship with our breasts. There is the first stage, before they come in, when we wonder if we will ever get them. Then once they do put in an appearance, we spend years worrying that they don't stack up, pardon the pun: that they are either too big, too small, too pointy, or too round, or that the nipples are too big, too brown, or (God forbid) have hairs coming out of them. By the time we get over worrying about whether they're perfect, they are inevitably imperfect, as time, gravity, and hungry little mouths have taken their toll. This brings on the "Should I or shouldn't I fix them?" question. Can I now have the breasts I always dreamed of? This was never a big issue for me, as I was busy thinking of the ass I always dreamed of. But for some women it is an important concern and this is the time when many women make a visit to the plastic surgeon for a boob job.

Regardless of how you handle the third stage, the fourth stage is the same for everyone. The last and final stage of a woman's relationship to her mammary glands boils down to "Please, just let them be cancer-free. Let me get them through this lifetime intact and healthy. I don't care if they hang down to my shoes, if they are covered with purple spots, or if they are flat like crepes. I just want them to be okay." It's amazing the different meanings "okay" takes on in relation to your breasts the older you get.

Breast cancer is clearly not going to go away, and while the incidence seems to be increasing, so are the life spans of people living with it. We have to do what we can to stay healthy, take care of our friends who are sometimes not so lucky, keep supporting the cause, buy the products whose proceeds end up being donated to research, and—despite the emotional turmoil it causes—show up for our annual mammogram and pray the odds are in our favor.

But—and it is a big but—there are a few things we can do to tip the odds in our favor.

One of my thirty-four doctors makes the claim that "up until fifty we get a hall pass from all the bad things we have done to our bodies, but come fifty the smoking, drinking, overeating, reckless eating, plus not exercising all catch up with you." While we don't have control over many things that happen to us as we age, we do have control over a few, including—and perhaps most importantly—how quickly and to what degree our body actually ages. Aging is very simply decay, unless you happen to be a steak or wine. The thing about decay is, we can't stop it, but we can control it—we can retard it, we can put it off, we can hold it at bay.

You have to accept that no matter what you do, some things—your eyesight among them—will start to deteriorate. Unless you happen to be a superhero or a freak of nature, once you hit forty the print gets smaller, and by fifty it all blurs together. But thanks to the new Lasik surgeries and advances in cataract operations, once you get to a certain point, you can reverse the decline. My eighty-two-year-old father does not wear glasses anymore. He had the cataract surgery and they put those lenses in his eyes and when we go out to dinner I need glasses to read the menu and he doesn't. In that case eighty is the new thirty and fifty is the new eighty. My eyes aren't bad enough yet to warrant the operation, but I would totally do it.

The number one thing every one of us can do to help ourselves look younger, feel younger, look healthier, *be* healthier, and enjoy the benefits of everything from more stamina to less depression is to exercise and lose weight.

Obesity and overweight are the number one contributors to heart disease, diabetes, arthritis, and endless other chronic health problems. Thirty-three percent of America is obese; this is a fact.

People tend to make the same two excuses for not exercising, the first one being "I don't have time." To this I say, you don't *not* have time. It's very simple. Make the time. If you have to get up an hour earlier, do it. If you have to miss lunch, do it. I like to remind people that Barack Obama exercised every day while on the campaign trail, and now that he's president, he still does. The busiest, most successful people I know make the time, and I think it makes them more productive. I would put money on that. The other excuse people give is, "It's too expensive! I can't afford to join a gym or take a class." Nobody is telling you to hire a personal trainer or spend twenty bucks a day on yoga. If you have the money and it's worth it to you, then do it, but one of the few things that is actually free in this world is exercise. What would it cost you to walk around the block? How about taking part in one of the many TV fitness programs?

You can spend $25 for a DVD and work out six days a week for a year at the cost of six and a half cents a day. If you can afford to buy this book, you can afford six and half cents a day to keep yourself in shape. And I'll tell you what: that initial $25 is a hell of a lot cheaper than a doctor's visit.

If you don't like to work out alone, get a group of friends and make a daily date to walk for a few miles. If you work, do it late or get up an hour early. You will add years to your life, and they will be productive, healthy ones.

If you read this book and walk away with nothing else, I would want it to be that you change your life and exercise.

Chris Crowley, author of *Younger Next Year: A Guide to Living Like You Are Fifty Until You Are Eighty*, points out that "every day your body makes a choice. It's either going to get a little older (decay) or it will get a little stronger." The only way it can get stronger is through exercise: strength training, endurance, and cardio work. You can, if not turn back the clock, at least make it stand still.

Let's face it: one of the problems with most exercise is that it's boring. And since the key to a successful exercise regimen is sticking with it, there are several crucial elements to consider when choosing your routine. The first one is finding something you actually enjoy. If you don't enjoy it, eventually (perhaps as soon as tomorrow) you will stop doing it. For decades I was a serial exerciser; I always did something, but I was completely promiscuous. I would give up yoga for Pilates and Pilates for strength training, and when I lived in California I would swim six months a year. Then six years ago I found the perfect exercise for me (emphasis on "for me"), but I have been doing it six days a week for six years and I have never gotten bored. I have dropped three jean sizes, upped my metabolism, and never been stronger.

What I do is called Core Fusion. It's a cross between Pilates, yoga, strength training, and a discipline called the Lotte Berk Method. In one hour you work out every part of your body, stretch, and strengthen, and for me it's as much a part of my life as breathing. I cannot imagine living without it, and I don't. Aside from the health benefits, working out is a great way to make friends. People in yoga classes become close; they go off on retreats together. It becomes not only a part of your life but a way of life. It becomes the way you choose to live. And as you get older, it is often all that stands between you and a walker.

The other major factor to consider in an ongoing fitness schedule is convenience. You may love that Thai boxing class across town, but unless your life is such that you can make it every day, it may have to be your weekend exercise treat. You have to find something that is either close to your home or office; if it's not, I promise, you will find legitimate excuse after legitimate excuse not to go. And eventually it will fall into the category of too much trouble and you will eliminate it altogether. My gym is half a block from my house; there can be a blizzard and I can get there. I have no excuses. You may not have the luxury of having a gym or place you like so close, but you *must* find something that is close enough that it does not feel like a hassle to get there.

Third, you must do it at a time that fits into your daily life. And make it nonnegotiable. Whatever the time may be, that is your time to exercise; it can change from day to day depending on your agenda, but I find most of the people I know who work out daily do so at the same time.

Think about it: every day your body makes a choice as to whether it's going to get a little stronger or a little weaker, and you have the power to decide which way it will go. In January, many people make New Year's resolutions to take off that extra twenty pounds, cut back on the booze, and exercise. So January sees the highest number of people running to their neighborhood gym and signing up. Everyone is turning over the proverbial new leaf. But do you know how many people actually follow through on their new plans? Very, very few. Forty percent of those who join health clubs stop going soon afterward, and 50 percent stop going within the first six months. Of those who actually continue, they go on an average of eighty-nine times a year. So while people may be signing up, for the most part they are not turning up. Joining a gym and going to one are two entirely different acts. One is loaded with good intentions, the other with vital results.

I am very hard-ass about this because I know, and all the research will tell you, that exercise is the only thing that works. Cardiologists will tell you, GPs will tell you, shrinks will tell you, anyone who knows anything about or cares about health, both physical and mental, will tell you it's one of the few defenses you actually have against looking and feeling like your grandparents' fifty, sixty, or seventy, and it gives you a chance to get to be an energetic eighty.

Exercise will keep your muscles from atrophying, your energy up, and your moods stable; it will decrease your body fat, up your metabolism, and strengthen your bones so you don't become one of those broken-hip stats. (My grandmother had genes on her side; she really shouldn't have lived as long as she did considering the care—or lack of care—she took of herself, but it was a hip fracture that ultimately brought her down. It's not that the fracture itself actually kills you, but it lays you out and makes way for a whole host of other things to invade your weakened body.)

I could write down all the things that can go wrong with us once we hit fifty, but our collective time is better spent addressing the ways we can feel better. The other no-brainer in this— and it goes with the exercise—is of course what and how we eat.

At this stage, especially for women, our metabolism slows down, even when we are exercising regularly. It's why all of a sudden women at around fifty have that little tool belt of flab around their middles. But besides the looks issue, which is not nothing, the old adage "You are what you eat" is, while important throughout your life, never more crucial than when you hit this stage of life. The thing about good eating habits is that habits are exactly that: the way you eat forever.

I do not believe diets work in the long run—and trust me, I have been on every one known to man. I spent twenty years of my life a good fifteen to twenty pounds overweight. I can still recite from memory the whole week of the Scarsdale Diet. I have done Atkins, South Beach, every fast. You name it—if it promised to drop weight quickly, I tried it. And sure, they all work: you lose ten pounds fast, then you go out and eat a tuna fish sandwich and six come back on before you've finished it.

When Bryant Gumbel was still on *Today* and his weight was fluctuating like the price of gold, he decided to do a survey of what diets worked the best and cost the least per pound lost. The only one that proved to have long-lasting results was Weight Watchers—and that is because it teaches you not how to lose a fast ten pounds but how to change your eating habits for life.

If you live in this country, you know how to eat. It is very simple. Even Subway and McDonald's are now forced to tell you the healthy choices. No one has an excuse. It's not like the fifties and sixties, when everybody smoked without knowing the harm it was causing. Starting in 1965, all cigarette packs had warnings on them, though they were not as ominous as they are now and people could still smoke everywhere including the doctor's office. Doctors smoked in front of you. Athletes advertised them. Think about it; it would be like Barry Bonds being the spokesperson for steroids today.

The same is true with food. When I was growing up, a few people—the ones my mother sought out, the Gayelord Hausers and Jack LaLannes—knew what could hurt you, but not many others cared or believed it anyway. But today we have nutrition information on every label in the supermarket and a proliferation of diet books, health books, healthy food books, and study after study about what is good and what is bad. It seems like every week they find a new berry lurking in the rain forest that is the antioxidant of all time.

It comes down to common sense: as intelligent Americans with access to the media, which reports on every medical and dietary advancement (sometimes even before they are discovered), we know what to do. I think there is something almost condescending in telling people what to eat and what not to eat. My eight-year-old knows. We all know. At this point it becomes like exercise—it's in your hands. We choose to do it, or not.

Everyone knows that french fries are bad and apples are good. Too much alcohol is bad; a little is okay. Meat every now and then is fine; a sixteen-ounce porterhouse every night is sure to do you in eventually. Sugar sucks no matter which way you cut it: it's bad for your weight, it's bad for your body, it's bad for your moods. It is a bad, bad substance. It's addictive and nasty, and I don't think it's a coincidence that it looks like cocaine. It's the cocaine of food. Soft drinks—stay away from all of them. Juices—the same. Sugar is the devil.

Diabetes kills more women than breast cancer. The numbers are staggering: between diagnosed cases, undiagnosed cases, and prediabetes, 80.6 million have it in some form. That is almost a third of the entire population. Twenty-three percent of those over sixty have it. But with lifestyle intervention you can reduce your risk by *58 percent* over three years.

In labs they grow sugar in tumors. A friend of mine who has been battling cancer for years told me that they tell you not to eat sugar in the two days before a PET scan, as it lights up the same way as the cancer cells. Does that tell you anything? The old notion that sugar was bad for your teeth was just the beginning; it's bad for everything. Look at kids who eat it and start getting hyperactive and uncontrollable. Well, it does the same for you, only you crash and get depressed, especially if you are menopausal and already suffering mood swings. And the thing about it is, once you take it out of your diet, you really don't miss it. I'm not saying you can never eat another cookie or dish of ice cream or whatever your favorite may be, but make it an occasional treat, something you indulge in every now and then. It's all about balance.

It's also about getting yourself checked out. Does everyone over fifty need thirty-five doctors? Probably not. Does everyone over fifty need to pay more attention to their bodies through regular testing for some basic things? Absolutely.

One of the biggest is to get your heart checked. While we are obsessed with our breasts and most of us get mammograms, women for the most part do not get their hearts checked until something goes wrong. It smacks a bit of the barn door, if you ask me. If you are over fifty, you should get your heart checked at least once, get a baseline, and figure out the rest of your plan from there. The point was really driven home to me earlier this year when I went in to visit my cardiologist.

Being in my early fifties and with zero risk factors, I often have people say to me, "Why are you going to a cardiologist?" Number one, I believe in getting checked. Number two, in this case I was going in for some elective surgery and needed a clean EKG to get clearance. I thought it would be a no-brainer; it turned into anything but, though I learned a lot.

I have a slightly irregular heartbeat, but no one ever worried much about it. Wouldn't you know—the day I went in for my EKG, my beats were all over the place. They looked like a three-year-old's drawing of the Alps. It didn't seem to worry the cardiologist that much, but I have this terrible habit of asking doctors so many questions that even if they think I am probably fine, after my ten-minute inquisition I convince them otherwise. So after I presented him with any number of possible disasters, he decided we had better get to the bottom of it and I should wear a heart monitor. Before I knew it I was down the street at the arrhythmia consultant (also known as an electrophysiologist), being hooked up to a twenty-four-hour heart monitor.

So around I went with my heart monitor, wondering why if

there really was nothing wrong with me, I looked liked someone hooked up to life support. Now, if you want to feel about
eighty years old and the least sexy person alive, look at yourself
with a heart monitor on; it's really pathetic. I hid under sweat-
ers and robes for the twenty-four hours I wore it, hoping my
kids wouldn't see it—and, frankly, trying to avoid looking at it
myself. But I managed to exercise in it and eventually got up
the courage to start talking about it, and the amazing thing is
once I started talking about it, it turns out all sorts of people had
worn them, were about to wear them, or were thinking of wearing one. I learned they are actually a good idea.

During the twenty-four hours you are supposed to just go
about your life; the doctor actually wants you to exercise, have
sex, and do whatever you do that gets your heart rate up. What
got my heart rate up was wearing the freaking monitor; it made
me a nervous wreck. And in terms of sex, please—I can't imagine how anyone would pull that off. The nanosecond the twenty-
four hours were up I ripped that sucker off so fast, I'm amazed I
didn't take a layer of skin with it.

Two days later I showed up at the office of my new electrophysiologist for my diagnosis. "Your irregular beats don't worry
me," she said, "because they come from the lower chamber. If
it came from the upper chamber, it would be something else,
something we might have to really investigate, but looking at all
your other stats . . . Don't drink too much coffee, and come see
me this time next year."

But of course I couldn't leave well enough alone.

"You're sure it's not my arteries?" She had not mentioned my arteries.

Well, that was all she had to hear. I'm sure what I was un-consciously saying and she was picking up on was, *What if you have left a stone unturned, and what if that stone leads to my having a heart attack?*

She said, "You're right, and your heart could be enlarged too."

Now, she hadn't said that before, and the cardiologist had not only given me an EKG but had done an echocardiogram, which shows if your heart is enlarged. I reminded the electrophysiolo-gist that the cardiologist had checked my heart and said it was normal size. "Well, not the back," she said. "He can't see the back and right side, only the front and left."

Great—my third heart test in as many days, and they still knew nothing. Talk about irregular beats! My heart sounded like Ringo Starr was practicing on it.

So she said, "You must go get an MRI."

An MRI? How the hell had we gotten from there to here?

"It's the only way to really find out," she said.

So now I had to go see another guy to get an MRI. Terrific. The day I went was snowy, my appointment was at the end of the day and they were running late, and it was my daughter's birthday. I kept thinking, *What if they find I have an enlarged heart and have six months to live? How will I ever sing "Happy Birthday" to her?* It was not the eighth or the sixteenth, and so I had no way of hedging my bets; I was left in the hands of fate.

At the MRI place I was the last one of the day. I answered

another nine-thousand-question survey. I checked off for the seventeenth time that week that no, I do not have epilepsy, seizures, or prostate cancer; I'm not pregnant, nursing, or suffering from kidney disease. I'm just a high-strung Jew! I actually think they should add that to those questionnaires. In-denial WASP? High-strung Jew? Accepting Buddhist? I always check off the same things, "asthma" and "anxiety," which I guess is as close to "high-strung Jew" as they can get without sounding anti-Semitic.

MRIs are not a day at the beach. I don't mind them as much as many people do; there was a period when I spent a lot of time in sensory deprivation tanks, and MRI machines kind of remind me of a scary version of those. But if you had your choice between a mani-pedi and an MRI, well . . .

This one required prep, as I would have two IVs so they could look at the arteries and the heart at the same time. Cool—one last test, kill two birds with one stone. *Deep breaths. I can't breathe. My heart is racing—oh, right, that's why I'm here.*

I got all geared up and climbed on the MRI machine. In came the MRI guy; it turned out he was a cardiac radiologist, he had gone to med school (I always thought they were like chiropractors). But he took one look at me and asked me about my asthma. I have normal asthma, I told him. What is normal? Mine.

"Well, in that case we can't do the test to see your arteries," he said, "as the solution we inject into your veins can have a bad effect on asthmatics. But let me go call the electrophysiologist to check."

So off goes the cardiac radiologist to powwow with the electrophysiologist, and I was thinking maybe they should have chatted before I arrived. But he was just being vigilant, and I didn't want to get my heart rate up. So I lay there listening to some machine that went *swoosh, swoosh, swoosh*, but if you listened really carefully, it sounded like it was saying *man down, man down, man down*—not a reassuring mantra under the circumstances.

He returned to say that the verdict was that he could do only the one test; they could not look at my arteries.

"But that's what I'm here for."

"Well, no, you are also here for a possible enlarged heart."

Couldn't everyone maybe stop bringing that up?

So into the MRI I went. Forty minutes of breathe in, breath out, hold your breath, and you think it's half over but find out it's only a third over. Then I started doing fractions, and at one point I pretended it was the sensory deprivation tank, but just as I relaxed it started making noises like I was on the inside of a Cuisinart. I started envisioning what I would look like if they had to crack open my chest, or who might donate a heart if I needed one.

Eventually they rolled me out, took out the needles, and told me to go get dressed so I could chat with the radiologist. On my way out I saw him sitting in the hallway looking at three monitors with my heart on them. I instantly asked him if I would live; he told me, "Yes, at least for the moment," and then told me to go get dressed. For the moment? He must have trained those girls over at the breast radiologist's.

After I was dressed, he patiently explained to me all the workings of my heart. It was kind of cool; I got to see it beating in a big photo. The upshot was that my heart is fine. It is a normal size, perfect. Great, I thought—I could get clearance for my procedure.

"Well, not so fast. The electrophysiologist is not going to give you clearance until she sees your arteries."

"Why can't you give me clearance? You're looking at it."

"I didn't order the tests."

When had the electrophysiologist become the head decision maker for my team? She was just the heart monitor person. This guy seemed totally competent, plus he had a full-blown picture of my heart in front of him. She only had some wonky inconclusive papers that had required her to send me to him. He was looking at the real deal.

"She ordered the test, she is in charge of the EKGs, and she has to give you the okay, which means we have to see your arteries."

"Do you think my arteries are okay?" I pleaded.

"I won't know until I see them. I imagine they are. But you know, I had a guy in here with perfect cholesterol, perfect blood pressure, and he was ninety percent blocked."

I hate it when doctors tell those stories, and they all have one: patient perfect who turns out to be patient almost dead.

"You can have a stress test," he said.

I didn't want a stress test. I knew I would fail a stress test. It's like taking a driving test again. It's one of the reasons I keep my

California driver's license, so I don't have to take the driving test again. I do not want to take tests I am bound to fail.

"If you don't want to take the stress test, we happen to have the gold-standard machine that will show you every artery and all the plaque buildup, and if it's okay, you won't have to take another one for years."

What kind of test was it? A CAT scan.

Swell—from the MRI to the CAT scan. But by then I was so convinced that I had heart disease, I decided I had better find out how bad it really was. "Okay, I'll do a CAT scan, but I need it now."

"Well, I don't have any technicians left." I realized it was almost seven and the office was closing.

"Tomorrow, then," I barked. "I have to have it tomorrow."

"We'll try to work you in."

I think this guy thought I was sort of demanding and it was better to get me in and out as quickly as possible.

At least I could go off to have dinner with the family knowing my heart was not enlarged and all I had left was to verify that my arteries were clear. Which at that point felt like saying, "All we have left is the Taliban to deal with."

This CAT scan was a breeze compared to the MRI. It was six minutes, another IV, but in and out. The upshot was I had zero plaque buildup. The "arteries of an eighteen-year-old," they told me. Now, this may not sound like much to you, but to put it in perspective, when I took the test results to the cardiologist's office, they said they had actually never seen a zero score in all the years they had been practicing. I had the lowest plaque buildup of anyone they had seen in over twenty years. This happens to be genetic; I cannot take credit for it, though I wish I could tell you there was something you could do to get that result.

So these tests not only alert you to when something could be wrong, like the guy with the perfect everything who was in fact a time bomb, but allow you and your doctors to know where you stand now, how you should move forward, and what if any medicines you should/could be taking to help prevent a nasty situation from arising in the future. And in my case, I now know that when I get myself all worked up and get the pain down my arm, get short of breath, and work myself into a tizzy, I'm not having a heart attack, I'm just doing what I do best—driving myself crazy.

5

If She's Fifty, Chances Are Alice Doesn't Work Here Anymore

There is no excuse
There is unemployment
There is no work
But you gotta find ways to work.
—DUSTIN HOFFMAN IN *Tootsie*

At thirty I got my first job working as a screenwriter for Walt Disney Studios; at forty-seven I was hired for my last job adapting a book for Warner Brothers Studios. I have not worked in the film business in any professional capacity since. By "professional," I mean no one has actually hired me to do anything. And what is really amazing is that as I write this, I have three films in the top thirty-five on iTunes.

There is something about this that is surprising yet completely predictable at the same time. In Hollywood thirty is considered eighty, especially where women are concerned. This attitude tends to affect actresses first, but the second group on its hit list is usually writers, particularly those who write comedy, a genre not very friendly to women to begin with. So, being a female comedy writer, I should not have found it so much of a shock when I suddenly found myself jobless, with few prospects in sight.

As in every profession, there are exceptions to the rule, and one of the biggest exceptions, if not the biggest, is if you are a superstar in your field by the time you are fifty, you can skid forward to at least sixty. This theory I have learned holds true in almost all professions. You have Diane Sawyer at sixtysomething, and she just got promoted. But she was a major star before she was forty. (And let's face it, she looks amazing. I think Mike Nichols must give her a formaldehyde dip every morning.) You can run down a list of women in their fifties and sixties in top jobs, but I promise you every one of them was a superstar in her world by no later than forty-five. The general consensus seems to be that if you haven't made it by then, the chances are you aren't going to, so why keep you around?

Nowhere is this more evident than in the youth-obsessed film business. For a screenwriter, "superstar" translates into pretty much one thing: how big is your box office? Box office is a place where size really matters. You can write the biggest piece of garbage of all time. You can have the most abysmal global reviews in the history of the . . . well, of the globe, but if you make money, if you get to a hundred million (provided the film didn't cost two hundred million), you're a star; at least in the eyes of those who write the checks, and those are the only ones who count.

There was a time when this wasn't the case. But in the last two decades they have kept raising the bar, to the point where it is now almost impossible for anyone to clear it. When I started out you just had to be good, turn in the pages, be reliable, and have a voice, and you could keep a career going for a good fifteen years. Which still took you to forty-five for many, but it was doable. Then there came a point in the early nineties, shortly after I started, when you *had* to get something made. It didn't have to make money; you just had to prove that you could get a film on the screen. Then they decided that wasn't proof that you had what it took, so around the time I had my first film made they decreed it had to make a certain amount of money, otherwise it didn't count. That amount of money has risen with each passing year. It's now to the point where you need to prove you can have an opening weekend that hits $30 million and muscles on to pass the $100 million mark to achieve real superstar status. Take it from me, without a superhero, a superstar, a stroke of fairy dust, or Harry Potter on your side, this is not easy to accomplish. But until you have proven that your movie is or is not going to do that, you can exist in the middle land between extraordinary success and life-shattering failure. While you are hovering in that unestablished zone, you can usually land another gig based on the "it might be the biggest thing since *Star Wars*" principle.

Film is not a business for the insecure, though oddly that

seems to be the dominant personality trait of its workforce. There is nothing the film business likes more than possibilities. It lives in the future tense. "It's going to be the biggest thing *since The Godfather*." "We are predicting a huge weekend." "There is no way there won't be a bidding war." "It should attract major talent." "This one is sure to be nominated for best picture of the year." Most things are qualified, and everything is always going to happen, about to happen, or on the verge of happening, because if you lived in the moment, you would have to face the cold, hard reality that most scripts never make it to the screen. The majority of people with promise never get the chance to realize it. And if you waited until things were a fait accompli, nine of out of ten times you'd be dealing with a failure of some major sort. So the only way to maintain the smoke and mirrors of a career and keep people's true state of mind borderline optimistic is by depending on what tomorrow may bring. This is doubly hard for a woman, as even under the best of conditions, tomorrow brings 75 percent less to them than it does to men.

The part about my story that is slightly confounding even to this day is that at the time my career came to what I consider a very premature end, I had just come off two big projects, one of which was expected to become a big hit. Every time someone mentioned it, it vibrated big box office, success, stardom for all involved—and that included me.

It was going to be my ticket to sixty, my $100 million baby, the film that was going to keep me on the train past the giant stop sign of fifty. It was the job I had been waiting all my career to land, and I got it just in the nick of time.

Through really hard work and single-mindedness, I landed the plum gig to adapt Sophie Kinsella's international bestseller, *Confessions of a Shopaholic*, for the ultimate producer (in terms of success), Jerry Bruckheimer.

Okay, granted, it's chick lit. But it's multi-million-dollar chick lit. In Hollywood-speak, it's multi-quadrant chick lit, which means it appeals to seven-year-olds and seventy-year-olds. And I am a chick—okay, a hen—but it was the right job for me.

How could this one not succeed? It was as close to a sure thing as you were going to find in a business where sure things don't exist.

I was set; when I wasn't in the shower giving my Golden Globe acceptance speech (films like that never get Oscars, but who cares?) I was decorating the house in southern India my DVD residuals would purchase, yelling cut in my directorial debut, or giving *Vogue* a tour through my well-designed closet. I was living so far into the rosy future that the impending catastrophe of the present wasn't penetrating any part of my daily life.

I did get to work on the script longer than any of the subsequent writers did. And I was told that while I had delivered "pretty much what they were going to shoot," they just wanted an "A-list writer on it for a few weeks." Translation: *someone more famous than you are.* And then they were going to shoot it.

So there I was, hovering in It Could Happen to Her–ville. This tenuous state of being practically on the verge of possibly finally making it really big allowed me to grab my next job. I was so on the brink of maybe being really hot that I was up for both *Gossip Girl* when it was going to be a film and the adaptation of *The Ivy Chronicles*, a lovely book about a thirtysomething woman who loses her job and has to take her kids out of private school, start over, and learn to live with less. Though I wanted *Gossip Girl*, as I felt it would be the bigger hit, Warner Brothers felt I was much better suited to tackle the story about a woman faced with downward mobility. How's that for writing on the proverbial wall? That was my last job, adapting a book about a woman who loses her job.

While I was doing that and someone younger was redoing *Shopaholic*, the vibe changed a bit. It does that out there. Hollywood tends to mimic the underlying San Andreas fault, shifting and moving all the time, always leaving its inhabitants on desperately shaky ground.

By the time I finished *Ivy*—which they also professed to like—I was forty-eight and on my way to forty-nine. I had been working for close to twenty years. The people who were running things were in their late twenties and early thirties; they had their own rules and their little clubs, and they preferred to hire their friends. For the most part they do not like hanging out with their mom's friends, much less working with them.

In Malcolm Gladwell's book *Outliers*, he talks about how it's not just having the talent or even the breaks that ultimately leads to big success; it is oftentimes the combination of extraordinary accumulation of experience (ten thousand hours of practice is what he says) but also when you were born. Thus on what side of the age divide you fall determines everything—from how much need the world has for your services to what kind of void has been left by the previous generation to what is happening globally. In terms of Internet success, he uses the cutoff point of 1955. If you were born in 1953, it was too early; 1958 was too late. Had Bill Gates been born the year I was, there is a very good chance that he would not be where he is today. In fact, Gladwell says he most likely would not.

Well, in terms of right place, right age, right time, what the market required, and what I had to offer, by the time I was approaching fifty, things could not have been worse. First off, like many industries, the film business started downsizing before Wall Street started sinking. When I began working in the early nineties there were so many jobs they basically just brought you into the studio and said, "So what do you want to do?" By 2006 the studios were cutting production way down; some studios, such as Disney, went from producing seventy films a year to twelve. Many of the independents went out of business altogether, and the foreign money dried up. Before they fell, the hedge fund types were the last guys throwing money at Hollywood; once they were taken down, it was an entirely new landscape. There were barely any jobs left.

And then add into the mix the whole reality TV revolution that wiped out hundreds of TV writers who had been making a really good living, including many who had become superstars before fifty. But even a superstar can be left jobless if there are no jobs to be had.

And despite the fact they are always shouting about how many women hold high-level jobs, show business remains one of the most sexist industries. Some cold hard facts: the percentage of working female writers in film hovers between 17 and 19 percent. And according to a Writers Guild of America study, that figure has not budged since 1999. Television is a shade better, but the numbers there are pretty uninspiring as well: only 26 percent of TV writers are women. In a report I read a few years ago, women were such a minority that they were nestled between Eskimos and Hispanics.

And then you have the ageism. Hollywood has always been famous for kicking out the old, and "older writers" are considered those who are over forty. Being in your mid-forties puts you smack dab in the land of older writer. So fifty would put you somewhere in the land of . . . ancient writer? Writer who has to be wheeled in with a defibrillator and don't forget to order the pureed carrots for his lunch?

And then comedy . . . well, that takes you plummeting to still lower depths. They feel women are much better suited for drama (our tendency toward hysteria, perhaps?), so your chances of employment in TV drama are slightly higher. But I happen to work in comedy and primarily in film, so without that megahit I had nowhere to go but down and, ultimately, out. Humor has always been a guys' gig, but never more so than now. There was a time when a certain type of female-generated comedy was in vogue, but that was a while ago and has been almost totally replaced by the *Pineapple Express/Superbad* genre, and I dare you to find me one woman who would ever get hired to work on those.

With the exception of forklift operator and coal miner, it would be hard to find such abysmal numbers for female employment. Truth is, based on gender alone, it's a miracle I got to forty-seven in this business.

So, sticking with Malcolm Gladwell's theory, I could not have been in a worse place at a worse time. Staying where I was would be difficult enough, not to mention moving up.

Despite the odds stacked against me, I continued to try. I came up with lame-ass boy movies I didn't even think I could deliver if someone actually said yes. I partnered with my best friend, the late Blake Snyder, thinking if I had some real balls in the room, they might get over my aging-girl status. I went in for the types of jobs I had been saying no to for years. I was at the point where I would do anything.

I pitched cop movies, soccer dad movies where the dad turns into a superhero, even talking-dog movies, a genre I had doggedly avoided for decades. If I was offered anything, I was in there giving it my all. I'm sure an air of desperation started entering the room along with me. And if they didn't know my age, I would announce it before they had a chance to notice. Then I would launch into how much I missed the old days, which only ended up making me sound like Sid Caesar reminiscing about *Your Show of Shows*.

One of my agents said I was getting bitter. Gee, I wonder how the fuck that happened.

This went on for a good eighteen months. I was rejected by everyone for everything, and twice by Queen Latifah. The only supposedly consoling factor was that it always came down to me and "one other person." This other person I always imagined to be the ubiquitous Josh (they are all called Josh): Harvard class of 1999, horn-rimmed glasses, interned for Letterman, followed by a stint at *The Colbert Report* and maybe one season on *The Simpsons*. One thing I promise you he wasn't was a menopausal woman.

By then I was a forty-nine-year-old woman desperately pitching a film called *Dan Parker: Attorney at Paw*. (The worst part of it is, I came up with that title myself.) And despite the fact that I almost gagged every time I sat down with my fake smile to try to convince some twenty-two-year-old that if she took this to the twenty-five-year-old who would then pass it on to the twenty-nine-year-old who somehow through the convoluted game of development telephone would get it to the forty-year-old head of production, who would inevitably pass, I kept on doing it.

Though I had been involved in this process for decades, this time really seemed to be the all-out worst. And not only that, I was genuinely befuddled by what I was being asked to do. It was like a professional chef standing in front of a stove and saying, "You want me to actually cook . . . with food?" I don't know if I was just at the end of my rope, knew my time in the industry was up, or was just so horrified by this particular project that every meeting felt like root canal. Yet I would manage to squeak out this fake little chuckle every time I stated the title, which was supposed to say to them, "Is this not the funniest thing you've ever heard?" They just stared blankly, as they inevitably had just seen *Pineapple Express*, which for their money was the funniest thing ever. I was getting old, at least for this game. And Dan Parker was the first guy to really let me know it.

I was so pissed off at myself and everyone around me—most of all the producer, Hal Lieberman, for never showing up to a meeting and leaving me to fend for myself with just Dan, the dogs, and the twenty-year-olds. I would storm from meeting to meeting, getting further pissed off. If people didn't dislike me before, they certainly did by the end of that go-round.

The final pitch was at Nickelodeon. I was sitting in a room with SpongeBob SquarePants staring down at me from every wall, trying to sell a dog movie to an executive whose face said he'd rather be playing Xbox. Despite these grim facts, as always, I did the best job I could trying to convince them of the "infinite possibilities" of this project. "Hell, Dan could even be a brand like SpongeBob." (Okay, desperate people say desperate things.) "After all, Dan has a law degree and dogs need defending and SpongeBob is just, well, a sponge." That did it. I had a moment of pure clarity: it was time to leave the party. There is a great John Patrick Shanley quote in which he talks about the point when your craft and your talent merge. Well, this was the point where my age, my gender, my years of experience, and my dignity merged.

I was from the generation who believed in real-life stories involving real-life people like lawyers with degrees, even if they were defending dogs, and I was selling to a generation who believed that a sponge who lived in a pineapple under the sea could be a blockbuster. And they were right and I was wrong. I don't speak cartoon.

It was time to get out. I was taking my marbles and going home. I would show them. The only problem is, they didn't care. They didn't want me. And it took me about a month to really grasp what that meant. For the first time in almost twenty years I was truly jobless.

I didn't like it. It felt weird; it made me sad. I needed to work. Perhaps Berlitz gave a course in cartoon-speak.

So I decided to renege—I wouldn't give up. I was a screenwriter, I wrote movies for the screen; that's how I made my living, that was my identity. It's what I told cab drivers when they asked me what I did. It's what I wrote on my visas when entering a foreign country. It's what I told the person sitting next to me on a train. What on earth would I be able to substitute for it? I couldn't let it go. I was putting on my flak jacket and going back into battle.

The problem was, nobody wanted anything to do with me. I couldn't get a meeting, a book, a script, a lunch. I was over—even though my agent kept saying I would have a job in three months, six months, twelve months. By fourteen months, no job had materialized.

I was coming to the end of the money I had earned from *The*

Ivy Chronicles. That's how you sort of fool yourself as a screen-writer; I believe this to be true in other jobs as well. As long as you have money remaining from the last job and you're not going into savings, you are not yet in the official sense "unemployed." I could still live in the land of possibility until the money ran out. Then it did. And then, for the first time in my career, there was nothing else in sight.

Once that money runs out and you look down the long dark tunnel of joblessness, the panic and desperation really set in. That is when I legitimately lost it. I would go to my office every day, the way I had for the previous decades, turn on my computer, and then plop my head on the keyboard and cry for hours on end.

Needless to say, this was not the most productive way to move ahead in the world, though it is a swell way to destroy key-boards.

This went on for months. I refused lunch dates, and if I had them, I moaned about not working. My career was over. What was I going to do? Pretty soon people stopped wanting to have lunch with me.

Those of you who have lost jobs know it's not only about the money (though the money is a big part, especially if you don't have a spouse to pick up the slack). It's also about "Who am I now?" And more important, when you get to the "third chapter," "second adulthood," or whatever euphemism you prefer for fifty, you are in the hardest demographic to get rehired. You have no idea how you are going to move forward. There are so many psychological hurdles to clear. Once you've made it over one, there is inevitably another, and then another and another. How do you get through the course? And once you do, who and what will you be at the end?

Losing your job, your identity, and your livelihood is up there with losing a spouse to death in terms of stress and strain. And not knowing how you'll replace it or match it or even come close as the years are ticking by makes for an even more devastating experience.

Many of the jobs that have been lost in the last five years are not coming back. And most of us either have accepted or will have to accept steep pay cuts. The best year I ever had financially was my last year in the business; now I'm making a quarter of what I once was. This is not an easy position to be in when you are putting kids through college, or if you happen to be one of the ones supporting ailing parents and trying to stash some away for your impending old age (which has almost impended).

After six months of daily crying, I finally decided it was time for a change. I had to do something. I had to move onward. I had to make a major life shift, and the only way I could do that was by working. But there was no work, at least not in the field I had been in, at least not for me.

I found a Virginia Woolf quote (and not because I was reading an abundance of her works in preparation for my own final dip in the river; it came up on Google's quotes of the day). But it made sense to me, so I wrote it on the whiteboard I used to use for my film outlines. It read: "Arrange whatever pieces come your way."

Arrange whatever pieces come your way. There it was, in big bold letters, staring me in the face every morning. It stopped me from crying, and it started me thinking. I had been arranging or attempting to arrange pieces that were no longer coming my way. I was scrounging around in my past and coming up with worn-out pieces that had seen their day and didn't fit into my world any longer. It's why nothing was working. I needed to find new pieces. I could arrange something. I could make something work, but it had to be with what I had at my disposal.

I think this is one of the giant lessons in adapting to age without losing your mind: we must let go of what was and begin accepting what is. And despite what they say, age is something that requires adaptation. That is one of the reasons this whole "fifty is the new thirty" nonsense gets me so irked. At thirty you have so many pieces to choose from, and if those pieces don't work, you have time to search out and arrange new ones. At fifty that is just not the case. You can find things to arrange, but you have to work with what comes your way or what exists in front of you. You can't wake up at fifty and decide you want to be a doctor or a lawyer; sure, you can go get your law degree, but can you imagine applying for a job as an associate at the age of fifty-five and then going up for partner at sixty? It's just not going to happen. If you are in a job where youth and appealing to the taste of youth are essential components, you had better start searching for some new pieces soon, since you will be the first to go. If you are in business for yourself, then no one can mess with you, but if it's a thriving business there is a good chance you started it years before. If you want to start your own business in your fifties, it's never too late, but I would start looking around for those pieces in your forties.

Like people who don't have a problem with menopause or looking older or any of the other side effects of the aging process, there are those for whom early retirement is welcome, in fact not only welcome but in many cases coveted, prepared for, and gloriously celebrated when it finally arrives. And to those people I say, great—Florida, here you come.

In the words of the ageless Simone de Beauvoir, "Retirement may be looked upon either as a prolonged holiday or as a rejection, a being thrown on the scrap heap." That sums it up pretty well for me. It's exactly what it feels like—giant rejection, being jettisoned to the scrap heap. I do not want to retire, not now, not at seventy. I want to retire when I'm dead. That's the day I plan to stop working.

And unfortunately for the people who do look at it as a prolonged vacation, the recession and the consequent demise of retirement funds have seen to it that this may not be a viable choice. So work we must, whether we want to or not—or whether they want us to or not.

Thus we find ourselves in a position in which the pieces may not be easy to find, but we must seek them out and arrange them so we can go on and be productive, so we can wake up every day with a purpose, so we feel useful regardless of what society may be echoing back. This is our job, and it can be done. I've seen it done.

"Arrange whatever pieces come your way." It's the perfect jumping-off point, and one I believe people in their early forties should start thinking about before they hit the half-century mark. You need to start thinking about and actually setting up some pieces that will be ready to arrange before you have to start scrambling around for them or find yourself left with difficult or unsatisfactory pieces. It's a process and it takes time and effort.

In describing someone who tripped himself up, a friend of mine aptly describes why fifty is so scary to both men and women: "As he was crossing fifty, he didn't really have time to turn things around anymore."

Time—how much time do you really have to turn things around at fifty? Not very much; that's the truth. Sometimes it's better to start from scratch and arrange those pieces that have come your way. Reinvention is possible in the "youth of old age." I've seen it done by many, including my mother and myself.

I think I was able to reinvent myself so quickly because I had watched my mother do it at sixty-five. She got to keep her job longer than most, but she had several things in her favor. She was a superstar at what she did; she was a big fish in a little pond. My mother was the gossip columnist for the *Santa Barbara News Press* for twenty-four years. She worked at home long before it was the norm. She worked her own hours, which tended to be at night, as she was a nocturnal creature. I can still hear her tap-tap-tapping away on her copy in the middle of the night.

She had very few rules, as the people who ran the paper were her friends and her column was one of the most popular things in the paper, if not the most. She had the freedom to write about what she wanted when she wanted, with no intervention from anyone. She usually ran in and left her copy on her editor's desk at whatever hour she felt like, provided it was there in time for the next edition. It was the perfect job for her, and she loved it. She got to meet everyone, travel the world, and call her own shots. I think she would still be doing it today, much like Liz Smith is at eighty-six, if she hadn't been "retired."

I spoke with Liz Smith shortly after she was let go by News Corp. She was devastated, and why wouldn't she be? She has the energy of ten people and still does a great job. She had a long run and she still has her syndication, but part of the reason she has the energy of ten people is she is still working, and God bless her. I want to be working at eighty-six.

Unfortunately for my mother, around the time she turned sixty-five, the paper she worked for was sold to the *New York Times*. The old guard, whom she had worked with for decades and who had allowed her the freedom she needed to do her job, was let go, and in came a band of young, go-get-'em journalists fresh off the college presses. They were going to do it their way, and their way did not include a woman in white gloves and covered in sunscreen, dropping off the copy of her choosing at 7:00 p.m. They knew they couldn't fire her on a whim. She was doing what she had always done, and she was doing it well. She had access to individuals most people never get access to. But they didn't care.

So they brought her in and said, "You can keep your job" (whew), "but there will be new rules. You will have to be at your desk in the office at nine a.m. each day." Her desk was at home, and if she got to it at 9:00 p.m., that was early. They told her she would get an hour for lunch and one coffee break. Her ideas for stories would have to be submitted to the staff at a weekly meeting, and they would approve which ones she could write. Her day would end with her clocking out at five.

They might as well have told her to go cut off both her arms and mail them to Baby Doc Duvalier to use as a centerpiece. This was not a woman who was going to clock in and out and hang her handbag over her chair while waiting for some thirty-year-old to give her permission to write. And I'm sure they counted on that being her response. She quit on the spot. They didn't fire her. They faced no age discrimination suit, yet they got exactly what they wanted. Like many in that situation, she felt the rejection of being thrown onto the scrap heap. I think deep down she knew she had had a great run; she had done it her way and well, and perhaps it was time to move on. But at sixty-five, where do you go? So she floundered for a spell. Most people in a similar situation do. She worked for a smaller, free paper, but it wasn't what she was used to. She needed to arrange new pieces, and luckily for her, pieces she had.

I have often found that the pieces that come your way are pieces that are already a part of your life. It doesn't always translate into walking by the drugstore and seeing a Help Wanted sign. Especially in the later years, you tend to want to do what genuinely appeals to you. Thus the pieces tend to come in the form of your passions, your hobbies, the things you truly love, or the things you have been putting off because you were too busy working at the thing that provided a good income. My mother, like myself, truly enjoyed what she did. She did not volunteer to stop any more than I wanted to stop writing scripts. But sometimes you really don't know what you are capable of until you are forced to find out.

My mother did not know how to do a lot of things; the only job she'd had since her twenties was as a writer. So she knew that somehow, some way, it had to involve writing. And she *loved* China. China was her passion. You cannot underestimate the power of having passions at this stage in your life and the role they can play in finding out what your next "job" may be. She had been traveling to China since the seventies. She had studied and collected Chinese textiles and studied them for decades. She had the world's largest collection of Chinese shoes for bound feet. I know it's odd, but it actually put her in a great place. She had cornered a market without intending to.

Her age was irrelevant; at sixty-five, she was in a position to do something few others could, and she had the foresight to see that. So she decided to sit down and write a book on shoes for bound feet. Now, you might wonder what the market is for such a book. I don't think she considered this at the time, or worried about how many copies it would sell; she was merely arranging her pieces. She is bright enough to know it wasn't going to out-sell James Patterson, but she was letting her passion guide her, she was doing something she knew more about than most, and she was making a niche for herself in the process. Plus she had a great time doing it.

Twelve years later the book is still in print. She has followed it with four others, and lectures around the world. I don't think she misses her old job one bit. At eighty-two, she is still active and working.

In the name of arranging your pieces you oftentimes have to let go of two things, one forever and one (hopefully) temporar-ily. Pay is numero uno—especially when you are starting from scratch or a form of scratch. Whether you have been a big earner, a medium earner, or a small earner, when you suddenly find yourself on Beauvoir's "scrap heap" you will undoubtedly be in for a reduced income. Scrap heaps don't tend to pay well. In fact, most second jobs don't pay what you are used to. Let's take it a step further: if you have been let go and you are fifty or over, the chances are not in your favor that you'll make the kind of money you once made. This is pure fact. For four years I haven't seen the money I saw in my thirties and forties, and I seriously doubt I ever will again. This is something it took me a long time to get used to. But you must—really. It is your new reality.

There are exceptions to every rule; some will fall onto the scrap heap with a golden parachute. I think the parachutes these days are more copper than gold, but the lucky ones will get sev-erance, and if they have worked for a big company with stock options, maybe the stock will be on the rise. Same with SEPs and IRAs and 401(k)s: they have taken a hit, and if you're not working you're not adding to them, but with some luck they will chug along and gather more value until you hit actual retirement age. Fifty is not retirement age, not for the vibrant, not for the engaged, and not for boomers who really have to work longer. That many of us have been forced into retirement due to global economic circumstances and ageism does not alter the fact that we need to work and want to work. Working very likely adds years to your life, and even if it doesn't, it makes the years you have much more satisfying.

If you are really lucky, you have a spouse who can support you, and your lifestyle will not change too much, because changing lifestyle along with changing everything else is a whole other burden that is heavy to bear. It may sound trite, but if you feel bad about yourself already *and* you can't afford to get your roots done, it really drags you even lower.

If you are the major earner in a couple or you are alone, you may have to take any job that will cover the bills while you are arranging the pieces that you hope will put you in a place where you are fulfilled financially as well as creatively and emotionally by what you do. And while it's indeed better to be in the position of having someone there to pick up the bills when you are no longer able, losing your job brings with it a great loss of freedom. People without the safety net may be saying, "Freedom, schmeedom—give me the cash." But women who have always made their own money have an independence and sense of self that is different from those who haven't, and it is very hard to let go of that. The ability to take care of yourself and to treat yourself or your kids without an explanation or request is a very powerful place to live, and losing it makes you feel very diminished in the world and very dependent.

I was with a friend the other evening, a woman in her late fifties who has been a big earner and had important jobs all her life. She now finds herself, like many, "consulting"—a legitimate occupation but almost always a second choice. For her I imagine it's a third or fourth choice. She is making a very small fraction of what she historically has. And while she has a wealthy husband, she hates not making her own money. It is a vital part of her identity. For her this has been one of the cruelest whacks of reaching her fifties. "Not making money makes me feel less than who I am, and I so liked who I was."

"Who I was" is the other big piece in the rearrangement process. It was one of the hardest parts for me. I had to let go of Tracey Jackson, constantly working screenwriter. I had hung so much of my self-worth as well as my income on this for so long, it was not easy to watch it disappear over the horizon. This was a moment I'd thought was still years away. I thought I would not only stay where I was but would move further up the food chain, that I had time to achieve my goals. But no—at fifty it was too late to turn it around.

I liked being a screenwriter. I liked being wanted and requested. I liked the adrenaline. And for all my complaining, I liked being in the ring. I liked winning. I liked seeing my name on posters and feeling like I had gone up against great odds and made it. I lost more times than I won, but I liked it. And I really, truly liked getting paid to do something I loved. I couldn't wait to get to my desk most days. To get paid for what you love is a terrific feeling.

You feel like you are at your peak after years of toil, like you have years ahead of you to put those skills to use, like you can earn enough money not just to live now but also to sock away some for your old age.

And then whack—you're out.

You really have only one choice: to arrange the pieces that are in front of you.

So how do you even set about the task of figuring them out, sorting through them, and finding the ones you can turn into gainful employment?

In my case I had relatively few options. We can start with the fact that I didn't go to college; though I'm not quite sure how much it would matter at fifty, I suspected it would take me out of the running for many jobs. And I'm past the age to be a Hooters hostess.

I'm a storyteller—that's what I know how to do. And I'm trained in the area of film. Like most writers, I wanted control of my material. But those aren't pieces, those are just wants and skills.

I wanted to make a film; I knew that. Nobody was going to give me one to make; I knew that too. So I would somehow have to arrange whatever pieces I had and try to make one. I had my passions: my children and India. Those are the two areas I know the most about and care the most about, and that's where I wanted to actually spend my days.

At the time I was dealing with a fifteen-year-old who was gong through the terrible teens. She was pretty much driving us all around the bend. Nothing earth-shattering, like being a junkie; just the daily wear and tear of angst and anger and taking her teenage problems out on us. She was like many teenagers: petulant and lazy, she talked back and was perpetually glued to her mobile device of the moment. She was not reinventing the wheel of teenage trouble, but it was trouble to us.

Over the course of the summer, I read a book by Madeline Levine called *The Price of Privilege*. A mini-hit, it was about how privileged children seemed to be the unhappiest, and it addressed many of the reasons why. The book resonated with me and reflected much that was going on in our home with Taylor. I started thinking if we were having these problems and Madeline's book was selling well, there was clearly a healthy percentage of families out there dealing with the same things we were.

Then fall came; it was Taylor's sophomore year, and things got worse. For close to a decade we have been involved with a slum school in Mumbai. It is run by a take-no-prisoners Canadian named Tania Spilchen. My husband and I used to always joke that what Taylor really needed was a month working in the slums with Tania; that would straighten her out. So what if I made a documentary about a spoiled kid (mine, and a major piece in front of me) who went to India to work in a slum school (another piece)? We would see if putting her in adverse circumstances would somehow show her how lucky she in fact was, and maybe she would return home with a smile on her face and a new lease on life. It had to be a documentary, as I couldn't afford actors or directors, and quite frankly, I couldn't even afford myself. I would make it short, shoot for three weeks over spring break, and call it *Spring Break*.

I had a project. Now the one big piece I was missing was funding. So I had to write a proposal and to look for some money. But once I started, I was busy: I had things to do. I was making a film. Small, but a start. I was back in the place where I couldn't wait to get to the office every day. I was able to raise some funds from friends and family, but mostly this was going to be a MasterCard production. Still, I was working, and at that moment, that was the only important thing. I had taken my pieces and I was arranging them, and the most amazing thing was they were falling into place.

What often happens when you finally hit upon the correct pieces is that other pieces you didn't count on start appearing out of nowhere. It was October when I decided to make the film. After a few weeks of arranging the pieces, I decided I needed to reread Madeline Levine's book. I looked everywhere in the house but couldn't find it. That night we were having dinner with two of our best friends, one of whom happens to work in publishing. He asked what I was doing, and I told him about the film I was making. (I was back to having a response to that question, as opposed to breaking into tears.) He told me I should get in touch with one of his writers—Madeline Levine, who'd written *The Price of Privilege*.

Talk about a piece landing in your lap. Arrange—arrange—arrange.

From that point on there was no stopping. The film about one teenager that I thought would take three weeks to make and would be called *Spring Break* turned into an eighteen-month shoot and resulted in a documentary called *Lucky Ducks*, which ended up being about teens and parents and how we raise our children and navigate these difficult years. It's about learning to set boundaries and what it means to let go. I wound up with 180 hours of footage. We traveled not only to Mumbai but to Marin County and Montana. I filmed experts on child rearing from many fields. It turned into one of the great creative experiences of my life.

I remember one morning at dawn I was shooting a scene in Mumbai. I had a crew of thirty (crews are very inexpensive in India) and a Steadicam following fourteen kids on the beach as the sun rose over the Arabian Sea, and I thought, *It doesn't get any better than this.* And you know what I did? I thanked Dan Parker, attorney at paw, for not coming to life, because if he had, I never would have found this other thing that I honestly love doing so much more than writing films that other people end up destroying.

I arranged my pieces and they rearranged me, and it was the best thing that could have happened. It taught me that I was and could be other things. That those other things could bring me pleasure and I could in fact be my own boss, something that is very important if I want to work for the next several decades. I learned that the paycheck and the glossy-sounding job did not my personality make. I learned I had strengths and talents I had never bothered to look for or tap into.

Gifts sometimes come in strange-looking packages, and sometimes they are hidden in the scrap heap. As in many other parts of our lives, the answers may well be right there sitting in our laps or sleeping in the next room or living inside our heads. You can make it work for you after fifty. It's not going to be what it was when you were thirty, but if you lay it out right and lower your expectations in the beginning, you might find yourself pleasantly surprised at the end.

So sit down—maybe in a room of your own—and meticulously arrange whatever pieces come your way. You will be amazed at what you come up with. Just start soon—don't be afraid. As Robert Lowell says, "fate loves the fearless." There is no better time to be fearless than in your fifties; if you really put your mind and your energy into it, you can make amazing things happen.

6

The Biggest Pink Slip
You Will Ever Get

A child enters your home and for the next twenty years makes
so much noise you can hardly stand it. The child departs,
leaving the house so silent you think you are going mad.
—JOHN ANDREW HOLMES

When I was thirty-two my first child was born; six
months into my fifty-first year she left home for college. It has for me been one of the hardest parts of hitting fifty.
I am not alone in this; every woman I talk to about it (and I
talk to a lot) feels a sense of sorrow, despondency, loneliness,
and at times even abandonment when a child leaves home. It's
one of the thresholds we cross that we are totally unprepared
for and I don't think we ever get over entirely. We may make
adjustments, as we have no choice, and we learn to live with
them and they become our life, but "adjustment" is the key
word in that scenario. And, sad to say, many parents never
adjust entirely.

As my father said to me recently, his mother referred to the "kids" coming over when she was in her seventies and he was in his fifties. Many parents never want their kids to leave, and if the children do leave, the parents want them back as often as possible. While some parents eventually turn their kids' room into a den with a fold-out sofa for when they come home—"You sure you can't stay through Sunday?"—many leave them exactly as they were the day the child left for college, BeeGees poster, swimming trophies, and stuffed animals all in their original places, waiting patiently for their owner's return. In fact, some parents leave their kid's room untouched for the next twenty or thirty years, or until the parents move out themselves. I remember seeing pictures of the late Dr. Randy Pausch's room that his mother had kept entirely intact—rocket ship wall paintings and all—and he was forty-seven by then. I suppose it's a feeble attempt to pretend that time has stood still and the kids will come home for more than a long weekend, Thanksgiving, or a week or two at Christmas or in the summer.

Our kids always remain our kids, and they are still children in our hearts. Somewhere in the not-so-evolved part of our minds we feel they belong at home. Ask most parents where their kid's home is and they will say home is where the parents are. Now, the kids may be living elsewhere; employment, love, and life may have moved them geographically elsewhere, sort of like Dorothy hurtling through the skies in *The Wizard of Oz*, but home is the place they lived before they moved out to start a life of their own. Everything else is just a house.

The exit of your children from the family abode is one of the first big stations we pass through once we've climbed aboard the Elderly Express. True, it's at the very beginning of the journey, but there's no denying that the journey has begun. The exception is if you had your children very young, in which case they could be heading off before your fortieth birthday. But with most of the people I know, their kids are taking off around the time they are turning fifty. It's a one-two punch, for women more so than men, and with everything else women are going through at the same time, it is often a one-two-three-four-five punch right in the gut.

"I still cry for days every time she returns to college," one mom has told me. "Nothing will ever be the same."

"The dog died the same week my youngest left for college, I truly feel like I am alone on an island," another friend writes.

"I keep his door shut so I don't find myself wandering into his room, sitting down at his desk, and crying," voices a mom in California.

"I pretty much cried her senior year when I was alone in my car. It helped to have the grieving begin months earlier before the actual dorm move-in."

"Grieving, alone on an island. Crying daily and then continuing it through the entire four years every time she leaves."

Does this sound like a bunch of hysterics who are unable to accept life on its terms? Are we needy, clingy women who are unable to acknowledge that time is marching on and our kids are at the front of the parade and we are moving to the back?

Did no one ever tell you that your children are merely on loan?

Come on, ladies, grow up—your kids do, and then guess what? They move out! They don't need you now. You had your time—you pushed the stroller and the swing and got to go trick-or-treating for eleven years. You kissed boo-boos and made cookies and got handmade cards that said "I love Mommy more than anyone in the world . . . my mommy is a princess." You got to sit through concerts, ballet recitals, swim meets, and endless soccer games. You watched really bad school productions of everything from Neil Simon to Ibsen. Admit it—you were there in the front row, elbowing your best friend out of the way so you could procure a better angle for your camcorder. And you made the costume too; I'll bet it's in a big box marked "Special Things." (Mine are.) You worried about colds, fevers, concussions, hurt feelings, and whether all those vaccinations really could cause autism. You dragged car seats onto airplanes, carried bikes on the top of cars, lugged camp trunks to buses. You were vomited on and pooped on, and later yelled at that you'd ruined their lives. You stayed up nights worrying about alcohol, cigarettes, and drunk drivers, and you checked for sex offenders who might be within an eighty-mile radius of your house. You bought first shoes, Tickle Me Elmo, baseball mitts, Barbies, bras, computers, mobile phones, Xbox and Wii consoles, video games you didn't approve of, and later on maybe even the pill or rubbers. You probably drove your kids more miles than a professional bus driver racks up in a twenty-five-year career. Basically, you had your time and your fun, so get over it. That is what "they"—the collective wise "they"—tell you to do. But take it from me: it is far easier said than done.

Unlike reinventing yourself for the workplace or dealing with the bags under your eyes or the tire around your middle, unlike finding new love or a new outlook, you cannot reraise your children. You cannot recapture the excitement of having them, the joy and the trouble at times in raising them. You cannot relive the wonder at their growth; you kind of have to wave good-bye, hope for the best, and pray that you turn into the kind of person they want to spend time with. Several of my older friends have said grandchildren are "bliss." But at fifty, the last thing I'm thinking about is being a grandma.

Seeing your children leave home is without question the universe's most brutal and unambiguous announcement yet that a lot of the really good stuff is now your history, that some of the most sublime moments in your life have been relegated to the past tense. There will be nothing quite like them in your future. You can never take those moments back, remake them, or relive them. It makes you feel older than a parent's death or your saddlebags can. I hate it as much as anything I have gone through thus far in adulthood.

People say you do get over it. But then other people say you don't. It is a case where everyone is different. Whether they walk away and never look back or stay connected, become your adversary or your friend, really depends on what kind of relationship you are able to forge with your child. I have seen good outcomes, and I have lived through and witnessed hideous ones. It's one of the parts in your life and future you unfortunately have little control over. The only control you actually have involves trying not to control your children, so they feel that time spent with you is not a burden but a gift. That part of the equation you have some power over.

Wanting babies, having babies, raising babies—it's so much a part of a woman's life. It is not only biologically preloaded into her hard drive, it's what motivates her into so many different forms of action throughout her younger and middle years. It causes her to make certain choices at certain times: the bad marriage when the ticking clock gets so loud she buys some earplugs to drown out the sound; pumping herself with all sorts of drugs when she can't get pregnant on her own; spending endless hours on eHarmony.com, Match.com—and, if it existed, PleaseGetMeABabyQuickly.com. The need to bear—and if not bear, at least raise—a child is probably the most urgent drive in most women's lives. Go ask almost any woman what the single most important event in her life has been, and almost always without missing a beat she will say the birth of her children (with the possible exception of my mother, who claimed it was meeting the queen of England).

Now with fertility drugs, surrogates, and other countries tossing adoptable babies our way like footballs, we can add a decade or two onto the time we have, if not to conceive and carry, then at least raise and nurture them. And then thanks to surrogacy, you have women in their fifties parading babies that a decade or so ago would have been their grandchildren but are now their children. Here is a place where fifty truly can be the new thirty. But in actuality it's not, because those women are fifty and not thirty, and in twenty years they will be seventy. Sometimes I want to stop them and say, "Excuse me, madam, but do you realize when that child is heading off for college you will be nearing seventy?" I never do say it, as I assume they have crunched the numbers. Though I seriously doubt they have any idea what they are in for.

As much as I miss my teen, I can't imagine being seventy pacing the halls and waiting for her to come home. I think that would be just too much at that stage. That is when you realize the universe's plans are not as wacky as we sometimes accuse them of being. The thing about kids, which is perhaps why we were originally supposed to have them in our teens, is they take enormous amount of energy. Without question I had much more patience and stamina with my oldest, who I had in my early thirties. And while I got pregnant in my early forties, I had much less patience with the actual act of being pregnant. I kept going to the doctor's office every other day in the final weeks and saying, "Take her out—I don't care how, just get her out of there!" At thirty I loved being pregnant; ten years later it was much more of a burden on my body.

Raising kids, like standing in line for hours to get into clubs, is a young person's game. But we want them; no, we crave them. We are genetically programmed to nurture, and, as we see with menopause, we are programmed to physically fall apart when

163

we are no longer able to procreate. On top of that, as I have learned from my own experience and that of my friends, psychologically we come unglued for a spell when our full-time nurturing duties end.

When our children march off to college and into their future as adults, our daily caretaking, mothering, and child-rearing duties are suddenly over. We have been essentially fired from the job we have been in training for, recruited to, and served in active duty for much of our lives. We are pink-slipped with no golden parachute to soften the blow. Many of us are truly devastated.

But how could we not be? We spend much of our early childhood playing with dolls in preparation. And we play with little dolls, dolls we dress and cuddle, dolls we rock and kiss, dolls that need us and fulfill our need to be caregivers. Find me a little kid who plays with a teenage doll and I will show you a future shrink.

Dolls do not grow up; they stay the same size until you decide when you are too old for them. Dolls never decide they are too old for you. They do not walk away; you do not leave them off at college, with extra-long sheets and a microwave of their own, only to find yourself alone in a train, plane, or automobile headed home to your empty nest. Little girls do not play in their Fisher-Price houses for a while and then downsize to a smaller one; they don't play coming home to a quiet, clean, childless plastic house with the doll in another town or state refusing to return to their nests. Nobody and nothing prepares a mother for what it is like when her children eventually leave home.

There is no such thing as a part-time mother. There are mothers who do not hold jobs outside the home and there are women who are employed by others, but either way mothering is a full-time job; some just happen to do something on the side that is either a career or brings in needed money. But you will still find them at home late at night correcting homework, organizing backpacks, filling out health forms, making play dates, and buying birthday gifts online. A mother never clocks out. So for the full-time mom or the working mom, the parenting workload remains pretty much the same, and in the emotional department it is identical. If anything, the working mom's emotional load tips over toward guilt, which adds an extra strain.

Mothering dominates and dictates your life for eighteen years. Nothing takes the constant vigilance and energy of raising children, whether you're a single parent or sharing duties with a partner. All mothers have the same fears and worries. And all have their lives and calendars filled with the same entries for close to two decades: sonogram, Mommy and Me, Gymboree, pickup at preschool, library day, dance class, soccer practice, birthday parties, holidays, plastic toy parts on the floor, cooking, favorite foods always on hand, clothes in the washer and in the dryer and on the table, "I can't find my sneakers," and "Can you pick me up at the mall?" There is something about lying next to a child who is falling asleep, about sitting in a dank gym watching your offspring win the game or fumble the ball, that gives you a sense that there is nowhere else in the world you should be. Few things provide this, at least in my experience.

Working moms (and I have always been a working mom) will tell you that no matter how much you enjoy your job or refuse to give it up, there is a little voice that says, "You could be spending more time with your kids." As is so often quoted, on their deathbed no one ever says they should have put in more time at the office.

Yet you wake one day and, without emotional warning, it's over, finished, kaput, never to return. Let me tell you, it breaks your heart in a way that is truly hard to understand until you go through it. And it usually happens around fifty. Just as everything else feels to be caving in around you, the rug you so carefully wove over the years is pulled out from under your feet.

I know, I know, I'm repeating myself. Just as I know they have to grow up and move on. It's only fair; you can't hold them back; it's selfish not to let them go. I know every one of the platitudes so well, I can recite them backward, and they are not without truth. But just as there is truth to death and sickness and love and divorce and everything else that is real in your life's journey, it doesn't mean it doesn't hurt like hell when you have actually lived the details of it. It doesn't mean it doesn't leave you breathless and sad and mourning for a spell. It doesn't mean you don't feel like your life and family will never be the same again.

I started preparing for my older daughter's exit years before she actually did. There were nights when I was furious over something stupid yet teenagey and annoying, the kind of irritation only a teenager flexing her newfound independence muscles can elicit. And oftentimes I would stand at her door, part of me not wanting to go in and show affection and thus validate her poor behavior, the other part fully aware that I had exactly 678 days left in which she would be sleeping in the next room and I would have the option to go in and kiss her goodnight. Having that option is a very valuable thing, and more often than not I would go in and kiss her. Sometimes if I was really mad I would wait until she was asleep and then lie next to her or just watch her and stroke her hair like I did when she was younger. I was always aware of the "on loan" fine print to the mother/daughter living arrangement.

I recently ran into a friend with her tiny infant. Without any provocation from me she looked down at him and said, "Imagine—someday he will grow up, go to college, and leave me." And they do.

I remember when I was well into my forties and going back to stay with my mother. I had been gone for decades by then, gone far longer than I had ever been there. I woke up in the middle of the night to see her standing in the doorway watching me sleep; her face was a mixture of joy and sadness. I was back in my room, in my bed (even though my bed was now a pullout sofa; it had taken her a good fifteen years to make that change), and even though I was a mother of two, on my second marriage, and deep in perimenopause, for her I was her child asleep in her bed in the next room where she belonged. She never said it, but her face did. I would be leaving the next morning to go back to my own life and my other family, but for that blissful moment I was in my bed in our house and all was right in her world.

I think I understood it better at that point than I had as a young adult, because as I lay there I knew I was headed for the same place. Soon I would be watching my child sleep in a bed that once had been hers but wasn't really anymore, as there was another bed in another room in another home that was her real home. I would become the punctuation in the story of my child's life; I would cease being the meat of the text.

The thing about this that makes it unbalanced is that for the parent the child is always the meat of their life's story. The child is supposed to go on and make her own life, of which her parents are only a small part. But parents, mothers especially, are tethered to their young, who become—even for working, successful women—perhaps the defining element of their time here on earth. It's really what they leave behind, and they are connected to them in a way that cannot be replicated in any other relationship. You can leave a marriage, you can lose a spouse to death and fall in love with another, but you cannot make more children, you cannot duplicate the experience, and you cannot replicate the attachment.

I remember once my mother saying to me—yelling at me, in fact—"I don't want to be the person you only meet for dinner." I thought at the time, *How silly. I will never be that person. I am your child. We will always be more.* But her relationship with her mother had been reduced to the obligatory meal, filled with small talk and little contact, and she did not want her own life to end up the same way. Thirty years later I was standing on Madison Avenue in New York waiting to have lunch with my daughter, who rushed up to me and greeted me with, "I have to meet Charles for lunch, so I'll just have a coffee with you." At that moment it all came flooding back, and I yelled, "I don't want to be the person you just have coffee with," then stormed off down the street. This was not one of my more mature moments, to say the least. But this was four weeks before she was to depart for college, and I was losing my footing as well as my emotional center. I didn't want to be the person she met occasionally for dinner; I wanted to remain the person and the place she came home to for dinner.

For the three months before she left for Boston I woke up every day with a profound sadness, my heart doing an internal countdown. Part of mothering was coming to an end. It didn't help that a month before she left, my best friend suddenly died, at the age of fifty-one. I felt like my life was melting all around me. Blake had always been in my life, and now he was gone. Taylor had been there for over eighteen years, and she was on her way out. Was this what getting older was all about? If so, I hadn't signed up for it, or maybe I hadn't bothered to read the fine print thoroughly enough.

As often happens with kids, Taylor got really good in the days before she left. The kid who'd spent much of her senior year avoiding me now could not get enough of my company. In fact, she clung to me like a baby koala. At her request I would lie next to her as she fell asleep the same way I had when she was young. To tell you the truth, it didn't make it any easier. It would have been better if she had returned to the snarky "don't tell me what to do," "who cares what time I come home, I'm in college now" person I had grown accustomed to. That person would have been far easier to say good-bye to. But she too was aware that part of our journey as mother and daughter was coming to an end. The daily to and fro, the constant presence, the absences that lasted for just hours and not months—all of that was coming to a screeching halt.

They say it's easier for the one leaving than the one staying behind. I think that is true. Leaving indicates forward momentum; staying behind is, well, staying in the same place. New worlds would be opening for her, and mine would bit by bit start closing up—not yet and not entirely, but this was certainly the beginning of what was to come.

I cried for months and months before she left. I would break into spontaneous tears when I just thought about it. Being the drama queen I am, one foggy summer day on Long Island I got in my car, turned on my sad-song playlist, drove out to the lighthouse, and just sat, staring at the Atlantic and trying to figure out where so much of my life had gone. Fifty years had disappeared. I knew where they were, I could recall them, but how did they add up to fifty? And hadn't she just been in first grade? Forget her—hadn't I just been a first-grader myself?

The last week she was at home was life in slow motion. The days were some of the longest I remember living through. Every moment somehow felt like the last something: the last time we order pizza on a Friday night with her living at home, the last time I wake up on August 24 in the rain and she's living at home, the last time I have to plunge the downstairs toilet with her living at home, the last time we go to Kmart with her living at home. And my overly dramatic nature only added to this never-ending dirge I was forcing us all to live with. It was clearly much worse for me than everyone else, or maybe I was just the loudest, but whatever the collective feelings, my runaway emotions made the whole experience much harder for everyone. Just as the steaks came in from the BBQ and I was bringing the salad to the table for a family dinner I would break into tears and moan, "Do you realize this is the last time we will all have porterhouses with Daddy's special sauce while Taylor is living at home?" My little one would say, "It's summer, she will be home in the summers, we will always eat steak." I was thinking in metaphors; she was living in reality. But reality was not serving my purpose at that time. My purpose was to mourn and be miserable, and as we all know, misery loves company, so I did my best to drag everyone into my hole with me. Toward the end of the summer even the dogs would run away when I entered the room. I suppose they expected me to start weeping, "Do you realize this is your last heartworm pill while Taylor still lives at home?"

For her part, Taylor was upset too, but like all kids off on a new adventure (or anyone off on a new adventure, for that matter), she was also excited—and scared and clingy and quiet and loquacious. It was truly a miserable week.

Plus it was confusing, because even though I would constantly declare that everything was the last, in the next breath I would add, "She isn't really going anywhere, it's only Boston, she can come home on the weekends." And I kept referring to it as "camp." I could not for the life of me say "college" until she was actually there. "Once Taylor goes to camp . . ." "You want blue or lavender sheets for camp?" I was truly losing my mind.

The final day I took my two girls down to the bay. We had ice cream for lunch, sat quietly, and looked out to sea. The reality was so evident that for the first time I didn't feel the need to editorialize the event. It was our last day all together with Taylor living at home. Or at least our last lunch, or at least for that summer. For me it was the last day I would be a mommy to this little girl, who was really a young woman. It felt like the end of a wonderful vacation in a destination I would never return to again; and truth be told, it was. In that moment of my misery, mourning, and drama I couldn't see there was a plane taking off to a new place. I just felt like my friend whose dog died the week her youngest left for school—abandoned on an island.

In that moment of many lasts, however, there was one first: it was the first time I had ever said that lunch would be made up of only ice cream. Lucy, my younger daughter, finally realized it must be a big event if I was dropping my rather strict dietary rules. But rules and departures be dammed, she was getting ice cream for lunch, so—true to her nature—she just enjoyed the moment. In my next life I'm coming back as Lucy.

Then the day came, we were packed and prepared—or at least we were packed. I had rehearsed the moment so many times. Would I break down on the street and cry or would I wait until we got into the cab headed for the airport? How on earth was I going to say good-bye to her? How were we going to fit all her things into the small room she would be sharing with someone else?

In July of that summer I had spent much of my time asking people about their experiences with letting go of their young—what it had meant to them and how they'd gotten through it. Everyone seemed to be slightly battle-scarred by the whole experience, even those who had been through it years before or had done it several times with different children. As is always the case when you ask, everyone had a story to tell, advice to give, red flags, and words of encouragement.

Our lives went on like that for quite a while; at least mine did. But as with most of the other big life changes that occur after fifty, you adjust. And beyond merely adjusting or accepting, your life is the life you have in the moment, not the one you had in the past. As several friends told me in my days of searching for some concrete answers to ephemeral questions, one day you wake up and you are just used to it. The great longing to turn back time and have a seven-year-old Taylor with me drifted into the past, and I didn't feel any older per se; I just felt different. And soon different wasn't so bad either.

As she adjusted to college and the petulant child slowly transformed into a version of who she is to become, we began to have more in common. Once she gained her independence she could stop fighting for it, and that released a huge pressure valve in our relationship. For her entire senior year, as I paced through the late hours of Saturday as it turned into Sunday, waiting for her to come home (like many mothers, I could not sleep until she was safely in her bed), I wondered how I would ever survive not knowing where she was once she was in college. How would I know if she was safely in her dorm or whether she'd been abducted or date-raped or God knows what? I would never sleep again, I thought. But person after person told me, in my endless attempts at having others carve out my future, that once they are gone, you really don't think about it anymore. You just let it go.

I said, "You mean you can sleep when they aren't at home?"

"Better," a few said. "You don't wake up all night and pace. You just somehow let it go and it's okay." And they were right. It was just like many years before, when I'd wondered how I would get through the days when she was first at preschool and I wouldn't know what she was doing and if she was okay. What if she fell off a swing, or got hit by a ball, or choked on a graham cracker? But just as all those fears somehow dissipated within a few days and the norm became mornings without her (soon to be followed by mornings and afternoons without her, and in the later years weekends where all I saw was the back of her head heading out the door), after the first week at college the norm became weeks without her. Now days go by where the only communication is in the form of brief texts, and I do in fact sleep through the nights. I was quite amazed by that myself.

Life is still entirely different without her. I miss the family of four; I would be lying if I said otherwise. I miss the idea that I will not get to play dolly with her ever again. I miss that those days are behind me, though when faced with the possibility of going though them again, I sort of cringe.

We had a brief moment a month or so after Taylor left when we pondered adopting an Asian baby. It wasn't the first time; a few years ago I wanted to adopt a baby from Mother Teresa's orphanage in Calcutta. There are hundreds of little girls there, all in need of homes, and I could have scooped up any one of them and brought her back here and loved her and started again. But the Indians are very strict in their adoption laws; they are adamant in their belief that fifty is not the new thirty. They don't even think forty-five is the new thirty. They won't let you adopt a baby if the mother is over forty. So I dropped that idea, or perhaps I should say that idea dropped me. But then one night we were walking down Bank Street and I was missing the fullness of the four-member family. A father with an adorable little Vietnamese girl walked by, I turned to my husband and we both said at the same time, "Why not?" All through dinner we played out the upsides such an undertaking would bring with it. We would give someone a better life, we could start all over with a new one, how much fun would we have, it would be good for Lucy, Taylor could get by with a smaller room because she wasn't there much at all, we are getting older but we still have a lot of energy—most days.

When we went home and told Lucy our potential plan, she had a fit. She swore to move out if any more kids arrived. She had shared us for years, and now was her time; she loved being an only child. While her opinions are important, I knew deep down she would adjust to a new sibling and learn to love her. But as I went to sleep I did a mental run-through of the reality of having a baby at this stage: *I am fifty-one. I have raised one child and am midway through another. I will be sixty when I pack Lucy up and send her to college. I'm already kind of old to be a mom. At this stage, do I really want to start in with the diapers and the strollers and the not sleeping and the never having a moment to myself and then kindergarten? And God, by then I'd be fifty-six.*

Lucy is now a tween, and she is fairly independent; over the next seven years she will get more so. My life will be less full of kid stuff, but it will be full of other things. The idea of adopting a baby was never mentioned again. I am in my opinion too old to be the mother of an infant. I am at a new stage in my life and I must adapt, and without realizing it I'd already started.

Everybody deals with the departure of a child in different ways. I know mothers who start new careers, who go back to school to learn that thing they never had time for. I know a busy mom who started taking Russian classes, since she always wanted to speak Russian. I know a mom who took up marathon running. There are lots of great ways of making new friends and new connections, of building new worlds that don't replace the old but fill the spaces the old ones have left empty. Me, I just get time to write more and spend more time with Lucy, and I appreciate every second because I know how soon she too will be off on her journey. And I do think I will start French classes in the spring.

I know many people who feel that they now get time to re-discover their husband, read more books, see more movies, and live a life that is not defined by numerous people's endless needs and schedules. Nothing takes the place of your children, but the truth is that life does move on and you do move with it. Your children are always your children, and they continue to need you throughout their lives in different ways.

Once I was able to gain some distance from the drama of the moment, I, like Lucy, could see it was not the last steak we would ever eat together or the last ice cream on the beach or the last Christmas or rainy day I would wake up with Taylor in the house. It only felt like it in those first months of the change. She is mine for life in one way or another. She is my child, and she will be connected to me until I die. Though our patterns of life will change, our connection will strengthen in some ways, and I will get used to the spaces in between the times I am with her.

And while our children do leave most of us in these very turbulent fifties, we still get some real mommy moments with them. When I went up to her college for parents' weekend, Halloween was just two weeks away and she wanted me to go with her and buy her costume. I was beyond thrilled to do it. I realized then how much there is ahead of us: weddings and her babies and jobs and a whole new life taking off. All that has its own thrill factor attached.

Then this other odd thing happened. Not only did we get used to daily life without her, but when she does come home the whole household goes a little topsy-turvy. Suddenly dirty glasses appear and towels are on the floor, the drawers in the kitchen are all open, and boxes of cereal are left out. I have heard similar things a lot from other parents too: the kids come and go as they please, you are not allowed to tell them anything, and they make their own rules even though they are in your house. You are now living with an adult (or a version thereof), not a child. You find that what you were mourning was the loss of the child; the adult is someone entirely new to get used to, and at some point you learn that adults do not belong at home with their parents any longer. Sure, you want them to come around and spend time with you, but they need their own places. Life does in fact move on, and when it happens and you have had time to adjust, you do. Beyond accepting it, moreover, you embrace it.

And from what I'm told, grandchildren are the best. You get all the fun of the kid, and just when they get difficult, you can return them.

7

You Don't Look Fifty

Women—they want to look beautiful.
—Fashion designer Valentino

When I was thirty I weighed twenty pounds more than I do today. I wore a size ten to twelve dress back then, whereas today I wear a six. My jeans were a thirty-one waist and now I can go as low as a twenty-seven depending on the cut. I don't state this to boast, though I would be lying if I said I wasn't proud that I have been able to pull it off at this stage. I share these facts because our looks are one of the few places where we can legitimately stave off the aging process. I think in many ways the whole "fifty is the new thirty" phenomenon stemmed from our ability to manipulate the physical aging process. Thanks to all the newly developed facial fillers, plastic surgery, and advances in cosmetics, plus our understanding of the importance of proper nutrition and exercise, we can look younger longer.

For the first time ever we can somehow alter almost every part of our physical being. We can fix crooked, yellowing teeth with either veneers or any one of a number of whitening agents out there. Do you remember when everyone with the exception of movie stars had normal-looking teeth? And by normal I mean a little stained, not ruler straight, and sometimes even with a minuscule chip. Now you look around and everyone has these mouths full of shining, perpendicular porcelain. It's almost creepy, until you realize that if it's not overdone it can make a difference.

And of course "Does she or doesn't she?" has been turned into "You're kidding—she really doesn't?" It feels or looks to me like most women of a certain age color their hair. And if they don't, they have many options to keep it looking thicker, glossier, and thus younger. So the Granny Clampett 'do, that horrible single-process witch look, or the old-lady-with-hair-barely-there look are no longer the only options.

There is no question that in terms of looks this is the youngest generation of older people to ever come along. Take someone like Barbara Walters—for God's sake, the woman is eighty years old! It's truly amazing. Go right now and Google her. Look at her pictures. The woman is a miracle. Now, granted, she is a miracle of medical science, but so are babies who are born at one pound and not only survive but grow up to become basketball players. I know she also spends *beaucoup* bucks and a colossal amount of time to look that way, but it can be done and she manages to do it and work at *eighty*!

Obviously women in the media need to look good. If Walters looked like my grandmother did at eighty, she would be cleaning the offices at ABC and not on camera. However, she is proof that in the looks department, eighty can be the new sixty. Looking younger than your chronological years can be accomplished to varying degrees, depending on how much time you have, how much money you want to spend, and how committed you are to the process. You can do it without a lot of money, but the time and commitment are mandatory. We can all look good. And I, like Valentino, believe all women do want to look, if not beautiful, at least really good, despite what they may say.

I know plenty of women who let themselves go when they reach fifty. They gain weight, their excuse being that it's beyond their control. They go gray for reasons I don't understand (though many women actually look quite good gray, most don't). They don't wear makeup or use products that might help their skin. They don't stay out of the sun, and most don't exercise. They pretty much do nothing to avert the havoc time can inflict on their physical being. As with taking HRT, surgery, injectables, and devotion to skin, hair, and body are choices people make. You actively decide you don't like the way you look and so you do something about it. Or you can give in to the whole thing and just degenerate like the women of yesteryear.

I find women who avoid taking good care of themselves all say the same things, like "These lines give me character." Really? I think most lines make you look old.

"I believe in aging gracefully." You can age gracefully while still maintaining a more youthful appearance. I'm not suggesting you wear short-shorts and a see-through blouse. A good haircut, a little makeup, and some tighter abs—I find a lot of gracefulness in that.

"It's the face God gave me and I'm not touching it." Really? God gave me a predisposition for a fat ass and thunder thighs, not to mention diabetes; I'm thrilled to keep them out of my life.

"Only self-obsessed, vain people spend all that time and money trying to look like a teenager." Did you know that all the female Supreme Court justices dye their hair? I rest my case.

And then there's the one that says you've totally thrown in the towel: "What's the point? I'm old." You are never too old to try. And if you say that, it only becomes a self-fulfilling prophecy.

People tell me, "You don't look fifty." Real people, not people I'm about to tip or who want something from me. I'm talking about people I meet on airplanes or new acquaintances. Like most people who hear that, I take it as a compliment, because what they are saying is that I don't look old. Despite all our yelling about how fifty is the new thirty, we still think of fifty as old. If it's old, then what does it look like? In most people's minds it still looks like my grandmother's fifty. In many cases, collective opinion has not caught up with modern science.

Our fifty does not have to look like our grandmothers' fifty unless we let it. Without intervention, the human body still degenerates at the same rate it did fifty years ago.

If I dressed like my grandmother, lived on cherry pie, didn't get my butt to the gym six days a week, and smeared Crisco on my face, I would probably look just like her. But I refuse to let that happen. Call me vain, call me shallow, but don't call me between eight and nine because I will be at the gym.

Why do I keep pushing the gym so much? At this point many of you are probably muttering, "Shut up about the gym already!" But I can't impress on you enough that when your body feels and looks good, it changes everything in your life. It actually makes you feel younger. You know the way old people get out of a chair, that sort of "one, two, three, plant both feet on the ground, hold on to the arms, and up we hopefully go"? How old do you think that makes you look, much less feel?

It also helps with depression. Let's face it, we've got enough going on that we don't need to deal with that too. And depression is one of the major side effects of estrogen loss. Exercise always lifts my mood, no matter how foul I may be feeling when I start out.

Exercise keeps you young, or at least as young as you can be at your age. I know seventy-five-year-old women at my gym who have better bodies than some of my nineteen-year-old daughter's friends. But the only way you get that is to work at it, every day. Like I said before: it's nonnegotiable.

Once your body looks good, you want your face to match. It's like getting a new sofa—suddenly the rug needs replacing.

The thing about starting down the road of plastic surgery or injectables is that many people have unrealistic ideas about how good they will actually look. They think that with a few nips and tucks they will turn into Christie Brinkley, even if it's a middle-aged version of her. The only way you end up looking like Christie Brinkley at fifty is if you looked like her at twenty. I didn't look like a fashion model in my youth, so how could I expect to pull that off now? But women play this mind game all the time. They say, "Look at Sophia Loren. She's had work done, and she is gorgeous." Well, sure; she is Sophia Loren, you moron!

If you take Sophia Loren, Cheryl Tiegs, or Diane Sawyer (nobody looks as good for her age as Diane Sawyer, but she was a beauty pageant winner in her teens) and do a few nips and tucks, a little Botox here, a shot of Restylane there, you're going to end up with a great look. (Now, that doesn't mean I'm saying any of them have necessarily done anything.)

I think many women believe plastic surgery is magic. They think they will go in looking like themselves and come out looking like Sharon Stone. Unless you start out looking like Sharon Stone, I promise you that you will not end up resembling her in the least. The best you can hope for is a better you. But sometimes a better you is better than a you who didn't make the effort to get better.

I have a friend, Joan Kron, who is actually the queen of plastic surgery; she wrote the book *Lift: Wanting, Fearing, and Having a Facelift*. When she was fifty, in 1978, she cowrote the book *High-Tech*, taking a little-known term and turning it into part of our daily language. At eighty-two, she is amazing. Google her; you will fall out of your chair, she looks so good. She is my role model: active, energetic, still writing and working for *Allure* magazine, she looks twenty years younger than she is. Who wouldn't want to be that? It can be done. But she too works at all of it. Unlike thirty, at fifty it doesn't happen without you putting in a lot of effort.

I must admit that when I was in my late thirties and early forties I took the "I'm going to age the way God intended me to" position. I don't know whether it was a hangover from my response to my mother's face-lift or something like my early anti-HRT stance. (Or even my "I'm breast-feeding all my children" proclamation—a pledge I did not make good on, either. When the time came I said, "I've been a hotel for nine months—I'm not turning into a twenty-four-hour-a-day restaurant for a year.") But I was very late to the "I'm going to do whatever I can to look younger" table, though once I got there I took a permanent seat.

I remember the first time I realized I looked old. Everyone remembers the first time. It's a milestone—it's like the first time you get your period or when you lose your virginity. It's that first heart-stopping moment when you look at your reflection and go, "Holy shit, who is that old lady standing next to me?"

For me it was on an escalator at Bloomingdale's department store in New York. I was between the first and second floors (how is that for time and place memory?). When I looked over I almost passed out: there was this really old-looking woman standing next to me wearing the exact same coat, purse, and boots. We had identical taste. *Good God, she was me!* For the first time I saw sagging jowls and thick droopy flesh around my nose. It looked like my face was melting. But how could that be? Just the day before I'd looked good, or a version of good. *It must be the lighting,* I decided. So I got off at the next floor and went running for another mirror. Nope—identical haggard, aged face sitting on top of my body. I zipped from mirror to mirror, floor to floor, only to be greeted by the same terrifying visage. It was in fact me, old Tracey, Tracey with a droopy, saggy face. I pulled the flesh up the way my mother used to do. *Better*. I dropped it back. *Help*.

Then it occurred to me: if I looked this bad in a department store, where they are known to have trick lighting that actually enhances your appearance, how bad must I look in the real world? Well, when I got out in the street and went home I realized: just as bad, if not worse. How had this happened overnight? And according to everyone I talk to, it seems to crop up overnight. It's sort of like when your roots go—they're fine on Tuesday, but Wednesday morning they are four shades darker.

I was forty-three at the time, already three years older than my mother was when she had her first face-lift. It was 2001, and Botox was still a year away from being approved by the FDA for cosmetic purposes. About a week later I was actually in a cab with my mother, who looked over at me and said, "Boy, do you need some work done."

So despite my protestations about never having a face-lift I took myself to a plastic surgeon for a consultation. Being a bit of my mother's daughter, I took my cues from *W* and *Vogue* and made an appointment with the doctor who was getting the most press at the moment. I wasn't convinced I wanted to do it, but I wanted to know what my options were. So I embarked on the first of many go-sees to many plastic surgeons.

It was everything you'd expect a top New York plastic surgeon's office to be. It was sleek and chic, and the women working there were in their thirties and looked perfect—well, they had all been redone by the man I was about to meet. I was ushered in and a slick-looking, well-dressed man with the air of someone who was getting a lot of press got right to the point: he wanted to know what was bothering me. I pointed to my jowls and sagging face. He looked up close and said, "There is nothing wrong with your face, but your eyes really need work."

"What? My eyes need work? What's wrong with my eyes?"

He pointed to the dark circles under my eyes.

"Oh, those. My eyes have always looked like that."

"Well, we can fix it," he said. "They're only going to get worse."

The thing with my eyes is that my entire family on my father's side has dark bags under their eyes. I know they get worse with age. His mother was referred to as Poopy until her eyes got so bad we began calling her Droopy. She eventually had them done, back in the seventies, and the guy, the doctor of the moment in LA, made a bad cut. As a result, one eye hung lower than the other, making half of her face even droopier and requiring her to have the whole thing redone. Another childhood face-lift memory that did not warm me up to the thought of being surgically altered.

I had been looking at myself with bags under my eyes for decades. I guess they had gotten worse, but it was such a slow process that I didn't really notice, not until the doctor of the hour pointed it out to me. He told me to forget the face for now; I was too young and it wasn't that bad. Eventually I would need it (well, sure, eventually I might need a pacemaker too), but for the time being, he told me, "I'd take care of the eyes, if I were you."

Great—I went in there worried about one thing and left worrying about another. My eyes had never bothered me before. Now I was obsessed with them and it all happened in about five minutes. Suggestion is a powerful thing.

I was then led to a frilly office where a perfectly pretty girl gave me the facts on how much an eye job would cost. It seemed like a lot then, but it seems cheap by today's standards. She then offered up a very detailed rundown of the recovery time and recovery process and the possible but improbable complications involved in such an operation. She almost had me until that moment, but once the words "recovery time," "recovery process," and "possible complications" entered the dialogue, I started to brainstorm alternative solutions. Sunglasses day and night for the rest of my life would be cheaper than the figure scribbled in front of me, and there was no recovery time involved and the only complications were if I lost them. (I don't like the idea of volunteering for anything that requires recovery time and includes a recovery process. It's what kept me out of these places up to this point.) So I took my pamphlet and all my info and the dates the good doctor had available and I filed them under "Plastic Surgery." That file would grow quite fat over the next eight years, while the bags under my eyes grew much worse.

This was right on the cusp of Botox taking hold of America's frown lines and freezing them straight. Today we think of Botox as just another cosmetic procedure; in 2007, 2,768,400 injections were given. Considering it has been on the market only for seven years, that is a huge number. To put it in context, in 2002, the year it was approved, the number hovered around 768,000.

When it first came on the market I knew a lot of people who became virtually addicted to it. I live in New York and work in Hollywood, probably the two most youth-oriented, vain cities in the country. Everyone I knew was going in and getting their foreheads and crow's feet obliterated. They were all talking about how easy it was, and suddenly everywhere I went everyone had perfectly smooth, creaseless foreheads and no nasty little lines darting around their eyes. But much like my militant position on HRT, I felt even more strongly about Botox. There was no way I was getting near the stuff; after all, it was made by the same bacteria that gave you botulism.

I remember when I first became aware of botulism. It was the canned soup incident of 1971. In that year a man died and his wife became very sick after eating a can of soup that was contaminated by botulinum toxin. It was a huge news story; anybody who was alive then remembers it. I had never even heard of the stuff, but suddenly we were all conscious of this poisonous substance that could kill you. It often lived in cans, and if you saw a can that was slightly bloated you were supposed to get rid of it, as it could have the deadly toxin inside. To this day I still dump cans that look a little wonky. When you talk about it with people they often remember the event but get the soup wrong—some say it was chicken noodle, others remember it as vegetable rice, and I always thought it was Campbell's tomato, but it was actually Bon Vivant vichyssoise. (I'm sure my mental mix-up is attributable to being a child of the Warhol generation, where all soup somehow morphs into Campbell's tomato.) But regardless of the brand and flavor, I knew botulism was dangerous. This is actually an understatement: according to the dictionary, the toxin produced by the bacterium *Clostridium botulinum* is "held to be the most toxic substance known to mankind."

And you want me to inject that into my face? For a second let's forget the idea of needles in the face—better than "recovery time," but not better than, say, lunch and a movie. This is the *most toxic substance known to mankind*. I wasn't getting near the stuff. And I held tight to that position for close to five years. I would go to parties and look at everyone around me; they all looked smooth and wrinkle-free, like they had been ironed out before they left the house. I had the crinkles and little creases that went with my age, but I told my husband and myself they were "character lines."

That is everybody's favorite euphemism for their wrinkles, which are said to give you "character." No—a good sense of humor gives you character, a quirky sensibility gives you character, the way you handle life's ups and downs gives you character. Wrinkles, crinkles, and lines in your face make you look old. Unless you're ninety-five or Georgia O'Keeffe, the average person just looks aged. And I was starting to in a big way. But I was immovable; I was not injecting myself with poison for the sake of mere vanity.

Here is the thing about these militant stands on certain, shall we say, aids or helpers: it's very easy to say "I won't touch a hormone" until you find yourself in the corner sobbing for no reason, sweating profusely, unable to sleep, have sex, or even get through a conversation. I was one of the women who was not going to get an epidural when I had my kids, I was going to have my children naturally—a position that lasted for about two contractions, at which point I would have swallowed botulinum toxin if they'd told me it would alleviate the pain. I love the women who say, "I want to experience the glory of the pain of childbirth." I say, "No, you don't—there is no glory in the pain of a nine-pound human being ripping through your insides and shooting out of your vagina." There is glory in the pain of the lyrics to a Leonard Cohen song, but not in giving birth; giving birth is messy, painful, and miraculous especially if you get through it without drugs. In my next life I think I will start the epidural when the pregnancy test comes back positive. But some people like it. Most, if they are honest, tell you it's beyond misery. So with many things it's very easy to take the high road until you find yourself miserable and desperate on the low one.

Somehow forty-three got to forty-seven very quickly and the lines got deeper. I would come home from evenings out and announce, "Did you notice I was the only person in the room who could move their forehead?" But my false pride was merely masking envy. I was looking older, and the eye bags were getting worse. I was still terrified of surgery, but the number of Botox users grew and none of the ones I knew had died. At that point my vanity overtook my fear and I resolved to stick one toe, or should I say a bit of my forehead, into the land of Botox. I would start with just a little, the smallest amount possible, to see how it was, like that first taste of spinach when you're a kid. If everyone likes it this much, maybe there's something to it.

While I was at my regular dermatologist's office getting my annual mole check I decided to let him shoot me up. Now, this is one cautious guy, I would not be surprised if Cautious is his middle name. He is not one of those doctors who in his off hours is shooting up his own face to the point of frozen maskdom; he looks more like a professor than a cosmetic doctor. The truth is, he really isn't a cosmetic doctor; they all now tend to do it out of necessity, but he comes from the old school and clearly had to learn new tricks. So he just did a little, and it hurt like hell. (It does the first time.) Once he shot me with the Botox, I figured in for a forehead, in for the jowls, and I had him shoot me up with a little bit of Restylane too. Now that really hurts, but I hadn't touched the jowls since I went to the guy who wanted to do my eyes, and they still bothered me a lot. In the intervening four years, multiple injectables had appeared on the market, one being Restylane, which puffs up the saggy baggies in your face. Why not try it?

With Botox, by the time you're home, you see results. In terms of instant gratification, it's an eleven out of ten. The Restylane left hard little bumps in my face and big bruises. I had been warned this could happen; even injectables come with a mini recovery time. I called Dr. Cautious when the bumps were still there the next morning. I was in a panic; I knew I shouldn't have done it. I had intended to go in for only a smidge of Botox, but no, I had to be an injectable pig and go for the Restylane too. He told me it was a normal reaction and to massage the bumps for a few days and they would go down. So I spent the next two days obsessively massaging my bumps, and eventually they did go down and my face in fact did look fuller and less droopy. He actually hadn't put that much in—they say you need one to three visits to achieve optimal results—but I certainly looked better, and when I went out I had the confidence of someone with a wrinkle-free forehead.

There was one small problem: every time I moved my head to the left I felt this tug, like someone was holding my skin while my head was going in the opposite direction. At first I figured it was a pulled muscle, but by the third week I thought something could be really wrong. It seemed a tad coincidental that it had started right after the treatments, so I called the doctor who'd injected me and told him the issue and he said, "No way"; he had never heard of such a thing.

If there is one thing I have learned in the last five years, it is that very few doctors will 'fess up to the fact that any of this stuff has side effects. A few will; there is a wonderful dermatologist in New York called Ellen Gendler who is totally up front about all of it and was the one who saved my face when the Sculptra fell, but most of these guys look at you like you're nuts even though I think they know exactly what's up. So if you venture down this road, be prepared for many people to pass the buck or make believe the buck doesn't even exist.

When the dermatologist told me it wasn't the Botox, I immediately leapfrogged to my favorite medical rock: "It must be a tumor." Now, I have the hypochondriacal quirk that everything is a tumor: a cold is a tumor, an earache is a tumor. I take the truly deranged position that all things are tumors until proven otherwise. So I went running to my GP. He asked if I had done anything to my face, I told him about the Botox, and he told me in no uncertain terms that was exactly what it was. The injection froze a muscle and it was stuck for the moment, but once the Botox wore off the feeling would go away, which it eventually did.

And once it wore off I wanted more. I was hooked. I was a Botox junkie. When I looked in the mirror I liked what I saw. After consulting with some friends who were living deep in the land of injectables, which is often advertised as "the results of plastic surgery without the recovery time," I realized the way to go is with a plastic surgeon. Look, you can get the stuff anywhere now; it went from behind closed doors to your neighborhood facialist shooting people up. But the truth is you have to be careful: they are injecting chemicals into your face. Ultimately, you want someone who understands the structural composition of the face. I'm not saying there are not plenty of qualified dermatologists out there using injectables and doing a swell job; after all, some were the first to use them. But I feel the advantage of a plastic surgeon is that he or she is someone who studied the face in med school—every muscle and fiber and bone. If someone is shooting me up with substances that are only recently FDA-approved—especially if some are the most toxic substances known to man—I want that person to be a specialist. Call me crazy. There is that rumor that dermatologists and chiropractors were the guys who couldn't get into med school. I think that is a little unfair, but while they studied skin, dermatologists did not study facial structure, and anyone who injects these substances into your face is working with the configuration of your face.

So my next visit was to another plastic surgeon. This one I picked, I'm embarrassed to say, because I ran into him at a party. I was not doing much homework on this at the time; that would change.

I had met this doctor once before, as he had done a lot of work on a friend of mine when she was in her thirties. By the time I ran into him he had made a name for himself and was toting around a fiancée who was done up to scary perfection. He too was slick and well dressed (must be a job requirement) and was known then as the "B and B" doctor—boobs and Botox. Who better to go to? Though I have never, ever for a second pondered messing with my tits, I am so in the minority, as breast augmentation outnumbers by a large percentage all other plastic surgery procedures performed in this country.

So off I went to see Dr. Boobs and Botox. The office was small for a plastic surgeon, especially one with such a large practice. I made it clear from the beginning that I was not shopping for surgery; I wanted fillers. I was a no-recovery-time type of girl. He was fine with that, but I think they all know that once they have you in the chair, you are a target. You may not be buying today, but as every good salesman knows, if you work it right you can over time convince someone they want what they say they don't. And the fact is I wanted it; I just didn't want it in the painful, drawn-out way. So I was easy prey; if he just stayed cool and let me get comfortable with him, before he knew it he would be scrubbed up and bellowing, "Scalpel!" and I would be in for some serious recovery time.

But that first visit was Botox only. And boy, did he Botox. He Botoxed my whole face—forty injections. I was amazed that I just let him do it. The thing about it is once you get that first shot, the others are like nothing—well, not entirely, they are like shots in your face—but you sort of get used to it. You go to the zone and live in the future place where you will look so fine. If anyone at any point in my life had told me I would volunteer for multiple injections in my face I would have had them certified as crazy, but there I was, getting shot after shot. I remember walking home and feeling kind of woozy. I had certainly crossed into the land of toxins quickly; I went from none to as much as my face would hold. I was sore and had little pinpricks all over my face, which totally creeped out my kids. I went to bed early, and when I woke up there was this evenness to my whole face. The thing Botox is so good at is relaxing you; it freezes your muscles into this peaceful state, and the little lines are immobilized somewhere in time, a time before you had them.

All was dandy. I was quite happy with the results from my visit to Dr. Boobs and Botox. I went off to have lunch with a friend who lives firmly in the camp of doing nothing to herself. She was an international beauty in her day, and is still a handsome woman. I think a little bit of something here and there and she would be back in true beauty status, but she is happy the way she is and quite vocal about her disgust at those who choose the other path. During the beginning of the meal I could feel water dripping down my chin. I didn't think much of it; maybe all the shots had left me a little numb. She didn't say anything, though it must have looked a tad odd. By the time pieces of salad were falling out of the left side of my mouth neither one of us could avoid it. I apologized as I pushed the beets back in, and muttered that I had had some Botox the day before. She said she'd been wondering why one side of my mouth was drooping. I said, "What?" With that, a whole mouthful of chopped romaine, cheddar, and avocado fell into my lap. I took out my compact, and sure as hell, one side was drooping off to the side. I didn't look younger; I looked like a stroke victim.

This woman is not one to be shy when it comes to "I told you so."

I ran from lunch back to Dr. Boobs and Botox's office. I was hysterical. He was rather blase; I was clearly not the first case of droopy mouth he had seen. The solution was simple: shoot in a little more and even them out. Why not? Was I making a pact with the devil here or what?

Perhaps I should have gone in for the recovery time. But by

the third day I was looking as smooth as smooth can be. Actually, I looked too smooth; there was something a little off about it. But I liked half of what he did, so when I went back three months later for a fill-up I had him do less.

In the meantime I had been making monthly visits to his office for something called dermal planing. These instant-result mini-remedies are truly the crack cocaine of cosmetology; it's hard to stop once you start. Every week there is a new miracle machine or molecule that will hurl you back to that place in time when you looked young and had more money because you weren't throwing it around on all these newfangled procedures. Dermal planing is actually a medical facial where they take a fine, almost razorlike apparatus and scrape off the dead skin, then do something with a machine and dry ice. It's one more drug to offer the beauty junkie, and since clearly I now had a full-fledged habit I was there for my fix. I would have my facials in a small room that in my mind was always the facial room; every now and then I would get a Botox shot or two, but it was the facial room to me.

One day after I had been going for about a year, he wandered in after my facial and said, "You know, you really should consider a lift, first of your eyes." I have to hand it to him; it took him a year to bring it up.

"I know about the eyes," I said.

But he felt, as many do, that while you are under the anesthesia you might as well get the whole thing done. And now was a great time: I wasn't too old, we could keep me looking this age or younger indefinitely—if we got to work on it soon. "If we got to work on it soon" made me sound like a coal mine that was going to cave in at any second.

Then he went through the procedure he would use; it was pretty standard. I listened, because it wasn't as if I hadn't thought about it from time to time. And I had been making regular visits for a year—how long had I thought he would hold off on this? He was there to operate, that was his day job, it's what he was trained to do.

I guess you could say I made progress in this conversation, since for the first time I got past the phrase "recovery time." I was forty-eight by then; there was no way I was going to get any younger-looking without some recovery time. It had taken me five years to adjust to the fact that I would eventually require some elective surgery that was going to be invasive and pain-ful enough that I would not be able to walk home right after and maybe stop at Duane Reade for my Tylenol. I would be packed in ice, sequestered from anyone's sight, and maybe need a private nurse to care for me. Okay, I got to that place, at least mentally.

He threw me a total curveball when he announced I would be "out for seven hours."

"Seven hours? You're only doing a face-lift; you can separate Siamese twins in seven hours!"

"We want to do it right," he responded.

"Yeah, well, certainly better than cutting off my nose and gouging out my eyes, but seven hours?"

I was almost out the door on that one, but then I reconsidered: maybe that's just how long it took. I had never discussed the length of the operation with the first doctor.

I was curious as to where he did this surgery. The first guy I'd gone to had showed me his operating room right away. The ones who have their own do that, because they're very proud of them; it's like showing off an in-house screening room. But where for the screening room it's "Here is my state-of-the-art screening room; it took four months to get the Dolby right," the plastic surgeon's version is "And here is our up-to-hospital-standards operating room. We have everything the hospitals do, minus the staph infections."

But Dr. Boobs and Botox said nothing of the sort. Without missing a beat, he said, "We do it here."

"Here in the facial room? You operate on me for seven hours here in the facial room?"

"Well, we get all the facial stuff out and turn it into an operating room. The chair goes back and we bring in the anesthesiologist, and frankly you wouldn't recognize it."

I would hope not.

"And tell me," I said, "after you turn the facial room into an operating room, do you turn the waiting room into a recovery room?"

"You recover here." No shit? What an all-purpose room—facial, operation, and recovery. I wondered: if I died in the chair, would they turn it into a casket as well? Not only did that whole conversation send me running out of his office, never to return, but I henceforth got dermal-planed elsewhere.

A few months later a well-known TV personality told me my eyes really needed work (think I was starting to get the idea?) and he would send me to his person, who sees no one without a referral. The TV personality said his person was the only one in the world I should go to. There seemed to be two things everyone agreed on: that my eyes sucked and that their person was the best. The person who told me this worked near the white-hot center of Hollywood beauty makers, so if anyone knew, he would. The fact that he had had so much done that it made him look like Cher was scary, but I figured, why not?

This person was not a plastic surgeon but a dermatologist who did surgery. (Yeah, I know I do many things I say I won't.) She wanted to harvest fat from my legs and shoot that into my face; that was when fat injections were all the rage. I have to give her credit for one thing, though: the most memorable line I ever heard in one of these consultations. She said, "Since you live here and work in LA and I have offices in both, we will store part of your fat here and the rest in LA. Your fat will be bicoastal." I've spent my entire life trying to lose weight, and now someone was offering me my fat on both coasts? She also wanted to do Thermage, a procedure that was briefly in vogue until the pain factor proved not worth the results for most people. Now you almost never hear of it. But it sounded worse than recovery time. And nobody was going to harvest my fat. My fat was staying right where it was, so I left yet another office with yet another folder and another total cost estimate, which would have been about the same as two complete face-lifts.

At this point I was feeling like I probably should have gone with the first doctor, who was the most widely recommended and offered up the least amount of work in the safest of settings, but by now the years they were a-passing. I'd gotten to the point where I was asking everyone I knew who did their work, who they thought was the best. I started really doing some homework. I have a few surfer friends who have had some gnarly accidents and needed their faces put back together, and for some reason that sounded like a good recommendation to me. By then I had been through two dermatologists and two plastic surgeons. This would be my fifth attempt at finding someone I trusted with my face, someone who sounded safe and reasonable.

I thought I'd found my man in the guy who ended up shooting me up with the Sculptra. I liked him instantly. He wasn't slick; he was like a funny lawyer or business manager. He seemed open and honest, and he had the best operating room in town, or so he said. He did a great couple of rounds of Botox and Restylane on me; I'd never looked as good and felt as comfortable with any of the others. It was like finally finding the person you're going to spend the rest of your life with. I was elated. I came home and told my husband, "I have found my man. This is the guy I will let put me under."

It was a glorious year we spent together, him shooting me up, me looking better and better and getting closer and closer to actually letting him attack my eyes. He gained much of my confidence one day when he was using me to show a younger doctor how to inject. It was from him I learned how much these guys, the surgeons, really know. He meticulously explained to the younger doctor, "If you go off a centimeter here, you hit the filbuoiusla minuialolea muscle, which controls the eyelids, and then the tivulia muscle will contract." I'm making up the names, obviously, but it was impressive med-speak that made me have faith that this man knew the physiology of a face.

But then I had the Sculptra-gone-bad incident. I have to take responsibility for allowing myself to get the stuff. I wanted it. It sounded good. I knew it wasn't widely used, but again, I did not do sufficient homework. Had I, I would have known it could lump up. It was primarily used in AIDS patients to fill out their faces when they get so drawn and gaunt. Sculptra has been approved for cosmetic use by the FDA, but not that many people inject it and most in the know tell you to stay away from it. There are many reported cases of people having to cut it out when it falls or lumps or both. It has a big tendency to lump up under all sorts of different circumstances. And when I ran into one of those circumstances, my trusted doctor told me he had never heard of such a thing. One more for the files—plus I was no closer to an eye lift, and by now I was fifty.

Eight years had passed since my first visit to a plastic surgeon. I had done nothing but spend money on injectables, cosmetics, and consultations, plus I had to buy a few more file folders to hold all the information collected at the various doctors' offices. The injectables work to a point, but they do not do what a face-lift does. Anybody really good will tell you that. Fillers do what their name implies: they fill the face and postpone the time when you will need to get a face-lift, but if you want really long-lasting results, eventually you have to sign up for the scalpel and some pain. My eyes had gotten really bad. There was nothing that any needle could do to help them.

They were so bad that whenever I was in a cosmetics department four salesgirls came running after me with the newest creams packed with everything from eye of newt to coffee berry, promising to remove all dark circles by the next day.

One day I'd finally had it, and I began to yell at them: "I know they're bad! There is not a cream, not a peptide, not an antioxidant in the world that will help!" Only surgery would do the trick.

But the truth is, I was scared. I still had visions of my mother with the tubes of blood jutting out of the top of her head. And then it seemed like every few weeks there were reports about plastic surgery gone wrong and how someone who'd just gone in for a little liposuction was now dead at forty-five.

Let's face it: this is not for everyone, though millions of American women have procedures done every year. In 2007 11.7 million women had some form of cosmetic enhancement procedure, and they spent $13.2 billion in the process. It is no longer a vanity ritual for the rich and famous. There are not enough celebrities to bring in that kind of money. How much can Heidi Montag and Jessica Simpson really contribute? No, the majority of the women who are doing all this spending are working women with kids, most of them in their forties and fifties.

So I finally gave in. I could not look myself in the eye with my eyes any longer. The good news was that I could afford it; I'd waited so long since the first visit that the money I had been stashing away in an "eye fund" had grown proportionately with the bags, and I had enough to cover the procedure.

And most important, I found the guy to do it. I'd actually found him about a year before; his name is Dr. Jon Turk, and he came to me through my friend Kelly Langberg, whom I met at— where else?—the gym.

Kelly had had work done by Jon and she looked really good— like herself, only better. In my ongoing survey of who had had work done and what they looked like, Kelly really stood out. So I went in to see Jon for Botox, which by then had become my litmus test: *If I like you and trust you and you don't bug me, I might let you do my eyes.*

What I liked about Jon was that he didn't push me. He didn't do the "Nice to meet you—come see my state-of-the-art operating room." Obviously, we discussed my eyes from the beginning, but he kicked back and just shot me up for about a year. In that year I saw other work he had done. And because he was so cool and had kept me looking natural, only better, with the shots, that made me want to do the rest.

I actually went from trepidation to something very near excitement. My fantasies moved away from the inevitable pain, recovery time, and possible complications and toward how I would look without Nana Dottie's droopy bags bringing my face down. I imagined what it would be like to take a stroll through the cosmetics section of a department store and not have fifty saleswomen jump out from behind their counters thrusting the latest concealer at me. I spent time thinking about what it would feel like to look five to seven years younger, to maybe not be so invisible anymore, to start feeling like I was not getting older and looking it day by day. I liked the way those scenarios felt; it all suddenly became worth the risk involved. The time had come to put to rest the long-held image of my mother with the blood and slits and see the reality that my mother's eyes at eighty-one were for the most part in better shape than mine at fifty-one.

I booked an appointment. Jon and his staff carefully went over all the details with me, and then I got truly excited. Oddly, once I was committed to having the procedure, the fear went away. I did the things you have to do to prepare, which is mostly to stop taking certain medications and drinking alcohol a few weeks before. Actually, the worst part was all the heart scans I had to go through to get clearance. But they proved to be a blessing, as they taught me how vigilant we all have to be with our heart and how easy it is to get it checked.

So eight years and several months after the first plastic surgeon told me I needed my eyes done, I woke up and walked over to the office of Dr. Jon Turk, MD, to have a blepharoplasty—that's what the guys who went to med school call an eye job.

There is something very comforting about Jon, so I wasn't afraid when I got there. And they let me keep my sweatpants on during the procedure. Don't ask me why, but something about keeping my sweatpants on really calmed me down. It felt like I was just getting a facial, only I was being put to sleep first.

The last thing I remember as the anesthesiologist was putting me under was someone asking what kind of red wine I liked. I mumbled, "Malbec," and woke up two hours later.

The first thing I heard was someone talking to my friend Kelly Langberg, who had sent me to Jon—she had called to find out how I was.

I was alive. I had had my eyes done and lived through it. And ... and ... I was not in pain—zero, zilch, nil. I could have walked out of the office, though they never let you when you've been under that long. But I went home in a cab, and after taking a nap for a few hours I woke up with gnarly-looking eyes but feeling fine. In fact, I felt so fine, I never took a painkiller stronger than extra-strength Tylenol, which I took for two days. They'd given me twenty Vicodin that remain untouched and in my safe, but I never hurt enough to need one. In fact, I never hurt at all. If I had known it was that easy, I would have done it years before.

Except years before, I had not met Jon, I had not saved up my eye lift stash, and I was not emotionally ready.

Recovery time is real, and it is longer than you think, or at least longer than I thought. My friend Kelly was out and about four days after a full face-lift. I stayed in my house for close to ten days after an eye lift. But along with the lift, I had a chemical peel, which is when they put a high-grade acid on your skin that basically burns off some of the old skin, then causes it to get red and crust over. Once the crusty bits fall off you are left with a new layer of lighter, wrinkle-free skin where the old, dark, crinkly stuff used to be. Sounds a little gross, but it's not bad and it's worth it. This takes longer to heal, as you are left with peeling skin and redness, and these cannot be hidden by makeup or a scarf, hat, and glasses, like the stitches and bruising from a lift. But you know what? The recovery time was actually my favorite part, aside from eventually looking better.

I planned it for the winter when I could take some time off. It snowed for two weeks, so going outside was not such a tempting thought anyway. I read, organized my closets, and watched movies, all unapologetically, as I was "recovering."

But once the healing takes place and you begin to see the results, which is pretty much two weeks later, you cannot believe the difference. I was—and I'm not making this up—a new person. Maybe some of it is in the mind, but enough was on my face that I knew I looked better, and that made me feel ten years younger. And consequently I acted ten years younger. It gave me a sense of confidence I had started to lack. I enjoyed putting on my makeup and looking in the mirror. Not in an obsessive-compulsive, narcissistic way, but in an "I'm not afraid of my reflection anymore" way. I was, thanks to Jon Turk, a new, better me. I was me ten years ago, or at least my eyes were, and they were the thing that had been really dragging me down. In terms of how I feel about the external me, it is without a doubt the smartest thing I have ever done.

And if you have had good work done, it doesn't look like you've had any work. (Good work really is different from bad work, and we have all seen the victims of bad work.) So people will say, "You look so rested," or "You've changed your hairstyle." And my favorite: no matter what you have done to improve yourself, when people notice a difference and they don't know what else to say, they always say, "You've lost weight." Me being the bigmouthed, honest person that I am, I always respond with, "No, I had my eyes done."

I do this for several reasons. One is that I want them to know. Two, I don't think it's fair for people to see you looking better and think it's just some fairy dust from above that happened to come your way but missed them. I can't stand it when people don't 'fess up. Let people know about what you had done, and inevitably, even if they will never get any work done themselves, they will admit to not liking something about the way they look. Many have been thinking about some part of their physical being that has been through the gravity Mixmaster and that they would love to see altered, and if you come clean with them, they will feel free to open up to you: "I hate my jowls." "I've been thinking about my upper lids." "I wish my thighs were different." "It's my ass." Something on all of us is not where it started, and the thought of returning it to its original spot is appealing to more people than just those who actually own up to it. At a certain point women do want to be, if not beautiful, at least closer to what they once were, and when they actually take the step and do something about it, if they are anything like me, it changes a big piece of their lives.

Some people argue that this is no way to chase down your self-esteem. They often go on to say, "One should find beauty within." Of course you should find beauty within, but you can have beauty within and also not feel horrible about yourself every time you look in the mirror. That argument just infuriates me. Why can't I be a good person on the inside because I want to look better on the outside? I had this discussion with a woman recently; she feels the problem with our society is our obsession with youth. Now, she is not wrong there, but that comes into play in areas that are so much more important to society as a whole than if some forty-five-year-old doesn't want to walk around with jowls. Let's start with the workplace and move on to dating and health care and openness about menopause and death and loss. If to get through this difficult stage in life someone gets a little boost by looking better than she does naturally, who the hell cares? If someone wants to feel better about herself and not cringe every time she catches a glimpse of her reflection, and medical science has made it possible, then why shouldn't she take advantage of it?

It's a narrow-minded and silly argument. You can be a decent, kind person who cares for others and the planet and not have to suffer the rest of your life with bags drooping down into your cheeks, or jowls that make you look like Santa. The indignities of aging are hard enough; if there is anything that can make it easier, I say go for it if you want to. And don't let anyone stop you. It's how *you* feel about you that counts.

8

I Didn't Mean to Spend It All

A penny saved is a penny earned.
—BENJAMIN FRANKLIN

At thirty I was lousy about managing money, and at fifty I have made little progress. Asking my advice on finances is like asking Michael Vick to come train your dog. But a book on aging and adjusting to the changes it brings needs to address finances, as they become a rather large issue the older you get.

I know certain things about money: you are supposed to save it so that by the time you are sixty you have a nest egg that will get you through old age. You can watch *Sesame Street* and get that piece of information. I don't have much of a nest egg or even a big nest—more like some twigs scattered about, and even those usually end up turning into handbags. I'm not proud of this. Nor do I advise it for anyone.

If I am lucky and get to live to a ripe old age, you might find me in Central Park surrounded by many expensive bags that once held what should have been my nest egg. I will be singing "We've Only Just Begun" to myself and will more than likely have a dog (not trained by Michael Vick; actually, not trained by anyone, as I'm as good at disciplining dogs as I am at saving money). Do say hi if you walk by; I will have a cup of coffee I can't afford and will share it with you.

I joke about this, but it is a legitimate fear among many women. "I'm going to be in a cold-water flat eating cat food" is something that goes through many women's minds, and for a very good reason. Most women will end up much worse off financially than when they started or were in their middle years. I can only imagine that with the recession, some portion of our nest eggs might have cracked. Like I said if you want real money advice I'm not your girl, go consult Suze Orman.

By the time they hit fifty many women end up divorced, some widowed; oftentimes the large Fisher-Price house becomes a smaller version purchased for a much lower price; some women end up leaving their Cleaver house for a rental apartment. A scary statistic from Chris Crowley, author of *Younger Next Year: Live Strong, Fit, and Sexy—Until You're Eighty and Beyond*: "Half the women in America approach the end of their lives at or near the poverty level." So the Central Park nightmare has a hunk of truth backing it up.

Once you hit fifty, your big moneymaking years may be either behind you or winding down. If they aren't and you are a good saver, move on to the next chapter. If you are not a good saver and are bad with money, meet me on the bench in thirty years.

There is still time, but you've got to get a move on. So do I.

Chris Crowley gives this very simple but spot-on advice: "Spend less than you make." You can also read *Confessions of a Shopaholic* and get the same advice. If you start now, you have a couple of decades to build up something, even if you are not making what you once were.

Living below your means in our society is hard, as we are constantly being bombarded with messages to buy more. Especially when you get to a certain age, you want nice things, you feel you deserve them; you might have gotten used to them. The media is forever cramming things down your throat: "If you buy this, you will look younger and hotter and your life will improve." We are often even more susceptible to these suggestions than teens are, as we figure it could be our last stab at any of it. Sometimes when I'm pondering buying something I don't need, I justify it with "How much longer will I really be able to pull this off? When I'm sixty I will not be wearing distressed jeans, a bikini, or a sequined T-shirt that says 'Vixen.'" (I don't actually wear a sequined T-shirt that says "Vixen.")

Or when you hear someone has just been diagnosed with cancer, you go right to the "It could be me next, so why shouldn't I have the latest bag, those expensive shoes, the super-duper face cream that will change my life just by opening the lid? I might be dead in a year." I have justified more inane purchases with this paranoid logic. Sometimes it's better just to go home, use the good china, or buy a lipstick.

Lipstick is proven to be the best splurge item for the money. It makes you feel good. It has a sexiness mixed with purpose, and you're not going to end up on the park bench in thirty years because you bought too many lipsticks. So I guess my financial advice is that if you have to buy something, if retail therapy is all that will ease whatever is bothering you, buy a lipstick.

The truth is, we really do have to start getting more serious about money in our fifties, no matter what our situation may be. For years I kept saying to myself that I would get serious and save. Then I woke up and I was fifty and I hadn't done much along those lines.

I have a pension with the Writers Guild that I can start collecting now. That is how old they think I am. I am pensionable. How did that happen? I'm so old now I should have a nest packed with eggs, as opposed to a closet packed with clothes.

It's not a coincidence that the scene from *Sex and the City* in which Carrie Bradshaw realizes that what she has spent on shoes is the equivalent of a down payment on an apartment has become one of the most iconic moments in the history of the show. Far too many of us relate; for the most part, we have been a group for whom instant gratification won over socking it away for a rainy day. (BTW, if you are fifty, it's getting cloudy.) The other famous Carrie Bradshaw–ism, "I like my money right where I can see it—hanging in my closet," has been too true for too many. The fact these characters have become so many women's heroes has done little to help the cause. Since it all works out for them (forget they get the clothes for free), it should all work out for us, but that is not always the case.

When I first went to work in TV I met a wonderful woman who gave me great advice. I unfortunately didn't pay enough attention, as I was young and rather arrogant at the time. Her name is Janis Hirsch; she suffered polio as a child, but she went on to pretty much conquer anything she set out to. She worked and still works in TV as comedy writer—not an easy job for anyone, especially a woman, and I imagine that for a woman with a physical disability it has to be even tougher. Comedy writers' rooms are not warm and fuzzy places. But she is talented and fearless and has stayed in those rooms longer than many of her male counterparts. Janis also went on to become a mother, something she was told she couldn't do. She is a force of nature, and I would love to see her now that she is in her fifties. But back when I was exactly thirty, she gave me two of the best pieces of advice anyone has ever offered me. She said, "Don't be a bitch; everyone is replaceable. And save half of everything you make." I wish I had listened, especially about the saving.

How many pairs of jeans, silly bracelets, and pairs of sandals that I "couldn't live without" have ended up in my yearly yard sale? My friend Barbara once said she had the down payment for a house in T-shirts alone, and this was long before Carrie Bradshaw uttered her famous shoe line.

They always say married women should stash some away in their own name, as you never know what could happen. It's a little paranoid and a lot sensible. I don't do it. My husband gives me money and I go out and get my hair blown dry and buy some face cream. I told you—don't do as I do.

If you are divorced and working and you don't have to support kids, living below your means is a great way to go. The *Tatler* once did an article that concluded the best way to save money was to avoid having kids. But if you are fifty, chances are it's too late to pick that route.

To be fair, some of our money problems have nothing to do with owning a closetful of Manolo Blahniks (I actually don't; they don't fit me), but if you have kids you know that they truly are ATMs in sneakers. They are constantly pushing your "withdraw cash" button. And now we are stuck with a group of them who, thanks to the dwindling job market, tend to need parental financial help long past college. Plus some people are helping aging parents or less fortunate siblings at the same time. As I said, it's not entirely our fault. And you may be living below your means already and still be unable to save.

We can justify our inability to save, as we come from the worst generation of savers of all time in this country. We need the newest iPod even if ours is fine. And the fact your cell phone or GPS or computer practically becomes last year's model even before you leave the store doesn't help matters.

What, no flat-screen TV? Who are you, Fred Flintstone? Get with the times. We have been a merry band of spenders and accumulators for years, and now we are facing that place where we should have enough or close to enough socked away to see us through old age. Most do not. I think that up there with breast cancer, being alone and poor is most fifty-year-old women's biggest projected fear, and it should be.

This I do know: you still have time. You have time to save. Time to learn to live with less. Time to revaluate your needs. Time to stash away that extra money your husband, boyfriend, or parents may toss your way.

If you are reinventing your work life, you must do so with your sixties and seventies in mind. Try to pick something you can keep on doing that will bring money in for many years, something you can keep doing even if your health isn't perfect. These are all scenarios we have to run through, as they are all possibilities.

The day will come for most people when we'll be living off what we have saved; we will have what we have at that point, and no more will be coming in. We are still used to thinking more is always on the way, so while it is, we really need to stash some extra away. You know how quickly it feels like we got from twenty to here? From what people tell me, the ride from here to eighty is much faster.

Our parents have Social Security. If you listen to the news (and it's always conflicting), there seems to be a good chance that we will be the first generation since 1935 who will not get it. If this comes to pass, it will certainly tarnish our golden years. Unfortunately, for the boomers, luck has not been entirely on our side, financially speaking. We got stuck with the biggest boom and the biggest bust since the Great Depression. Just as we were hitting fifty the jobs started to go, the market went to hell, and houses lost their value. Many things we were counting on to pull us through have diminished, if not vanished. It's all for real, and we have to deal with it.

So even the Carrie Bradshaws are making real attempts to get serious. It's late, but not too late. If we wait another decade, it will be way too late.

Try living with less; it is sometimes very freeing. I subscribe to a blog called Zen Habits that offers daily advice on how to strip down to just what you need. I try to follow it. Some days are better than others. We moved recently and I unloaded a ton of stuff; we literally now have only what we use, and I love it. You feel lighter of spirit. (Plus selling off things is a way of bringing in some cash.) If we buy something new, we get rid of something we already have.

I've tried to cut down on shopping, too—and please understand, I *love* to shop. I am the poster child for boomer consumption; I am not proud of it, just honest. But about ten years ago, in an attempt to get serious about saving, I invoked the twenty-four hour rule: if I saw something I liked, I waited twenty-four hours. If it was still on my mind in a significant way, I then bought it. After the recession began I upped it to the forty-eight-hour rule. And now in my fifties I have turned it into the seventy-four-hour rule. Pretty soon I will wait three months and by then it will be out of season.

Sometimes my older daughter, who has no concept of these things (consider her role model), will say, "What if it's not there in seventy-four hours?" Then I take the philosophical view: if it's not there, we're not meant to have it.

Because the topic of money management is so far from my field of expertise, I asked an authority, Mary Caraccioli, who has an MBA and is an Emmy Award–winning financial journalist. She generously shared with me some of her tips on how women should start dealing with their finances from forty on.

Basically, you have to quit waiting for the lottery, a miracle, or—God forbid—Prince Charming. They aren't coming. For all our gender's accomplishments during the boomer era, it puzzles me that smart, educated, accomplished women are still waiting on someone else to make their financial dreams come true. Wake up and smell the double espresso, then get to work. There is no magic to money. It takes the same elbow grease that goes into planning a great vacation, a fabulous party, or a stunning renovation. If you've done one of those three, then you can lead the money aspect of your own life. Once you take ownership of your financial future and give up on the lottery dream, you can actually build wealth or at least stop the bleeding of your cash for trivial things that don't give you what you really want.

How do you make that happen?

1. Do a little work on yourself. Understand your emotional attachments to money. What does money represent to you—status, safety, greed? Who helped shape that belief—parents, an ex, your favorite aunt? Whose money lifestyle do you most identify with now—who do you think is doing it right? Once you know your emotional ties with money and your financial

role models, you can more easily spot your triggers. This self-awareness has to be factored in when you build a money plan for your life. Knowing yourself means you can create a realistic plan that honors where you are now and where you want to be in ten or fifteen years. If you do this, you'll be ready to create a plan of what you really want in the years ahead. Make sure you write it down.

2. Buy a gorgeous filing cabinet, one you really like and won't mind using. You are going to be spending quality time with that cabinet and the files you are about to put in it. Carve out four hours a week (preferably in one or two chunks) and organize or reorganize all of your life's paperwork. Go through the account statements and the health care documents, shred the old stuff that has no value (such as cell phone statements from the last decade), and keep the stuff you need (stock purchase information, ten years of tax info, etc.). Then figure out what you are missing. Do you have info on all accounts—pensions, 401(k)s, investments, bank accounts, and so on? Are the beneficiaries for these accounts correct and updated? Even bank accounts should have someone named as the "payable upon death" recipient. Do you have a contact sheet for important advisors such as your CPA or tax accountant, financial planner, lawyer, executor of your estate, and so forth? Do you have a will, a living trust, a health care proxy?

You have to have a will. And if you are over fifty, when was the last time you read it? Is it still valid? Do you need more than a will now? At this point in you life you have accumulated something of value in your life, even if your checkbook is constantly in the red. Read up on estate planning. If you have kids or property and you don't have a will, you have to get your act together. Just Google "will" and you will find out everything you need to know about one. Don't wait on this—you could step in front of a speeding bus tomorrow, and if that happens and you have no will, a judge who knows nothing about you will determine the guardianship of your kids and who gets what from your estate. If you don't have kids and you care who gets your bobblehead collection and the beach house, then you have to have a will to make sure your wishes are granted.

Make sure you have a health care proxy. This little document will designate who will make medical decisions for you in case you are unable to speak for yourself.

Yes, I know this stuff is morbid, but do it. You will actually walk a little taller once you get this part of your life in order.

3. Make peace with your 401(k). Yes, just like the skin above your elbow, your 401(k) isn't what it used to be. It's sagging with almost no hope of bouncing back. The reality is that the stock market will very likely grow more modestly in the decade ahead. It's the hangover from a bubble party, from techs to homes and every asset class in between, that grew to enormous proportions and then burst in the last decade. Today you need to know what you have and guesstimate what you will need. There are a million retirement calculators on the Web; chances are your 401(k) plan has one on its site. Try it and assume very modest growth, 3 percent after taxes. Once you get a realistic idea of your retire-

ment income, then you can make a better decision about how long you need to work and what kind of lifestyle you will really be able to afford later. It may be a little bit of a downer, but remember we are checking our fantasies at the door. You want a roof over your head in retirement, not a park bench. No, it may not be a condo in Boca, but get over it. Too many of our mothers are living in poverty. Don't add to the statistic because you didn't plan.

If you have not saved a penny for retirement, that is a bummer, but your situation is not hopeless. You need to get your plan together. Social Security may or may not be there for you—and even if it is, it's not a lot. What kind of work can you do now to start contributing to savings? It may be worth getting a second job or consult on the side so that you can start stashing some cash away. If you find yourself in this category, pay particular attention to your health. Stay on top of things and lose a few pounds. You want to go into fighter mode; think of Linda Hamilton in *Terminator*. Get ready to take on the enemy, which is time. You want more of it to build a nest egg for yourself. If you are healthy and strong, you can do it. Other women are doing it; so can you.

4. Don't look for a miracle in the market or you will get a Madoff. If you just read the last tip and you think you will do better than others in the market or can get some sort of magic return on your investment—snap out of it! It is not going to happen and thousands of charlatans wearing suits and nice smiles are waiting to rip you off. They will be recommended,

like Bernie was. Magic returns don't exist. Someone offering to fix a complicated problem in one easy step is lying. Don't fall for it and please don't even look for it.

5. Pay attention to taxes. This could be a hidden strategy for making your money last longer. Understanding how your retirement income or your estate will be taxed is powerful stuff. The lingo is totally boring, but the wealth it can preserve is glorious. Get a good CPA. You want someone who will be in business for a while, so don't be afraid to hire younger. Having a trusted accountant retire on you just when you need him or her most stinks, but it happens all the time when you hire someone in your age group. Don't let an accountant manage your money. I believe in a true separation of church and state here. You want someone who is giving you tax advice, period. If you want a financial advisor, don't let that person manage your money either. Pay the advisor by the hour to give you sound advice about where your money should be—period. In this day and age, open up your own discount trading account and buy your own index funds or ETFs. That is likely what a good advisor would recommend anyway.

That's what someone who actually knows what she's talking about has to say. The only problem I'm having is deciding what kind of filing cabinet to get, and since they last a long time, I should probably invest in a really nice one!

Seriously—and this is serious—it's amazing how little we really need to be happy. From what I hear and see, there are three things that make the transition from the youth of old age

into the old age of old age pleasurable: being somewhat financially secure, having your health, and having some form of companionship. The money part is the one part we can actually
make happen. We can work on the health, and we can pray for
the companionship.

Benjamin Franklin lived to be eighty-four and died a wealthy
man. He clearly followed his own advice; at this stage we would
be wise to do so as well. But what confuses me is when you look
at him, you see that the guy also had great shoes. I wish he were
around to tell us how he pulled that one off.

9

Maddening Men

In olden times, sacrifices were made at the altar,
a practice that still continues.
—HELEN ROWLAND

At thirty I was coming to the end of my exasperating decade of dating and trying to find a mate; at fifty-one I am happily married to my second husband. I went seamlessly from one marriage to the other. Well, actually, there were plenty of seams. In fact, there were rips and tears—two divorces, moves across the country, breaking up two homes, adultery, and many hurt feelings among four people. But it felt like a small price to pay in order to avoid dating in between. That means I have not actually been on the market or in the bars—now it would be online—in twenty years. So I have as much knowledge about dating as I do about money. Except I eventually got better with the dating.

This is what I remember about dating: it was fun in my teens, as it was new and one of the first really grown-up experiences I got to take part in. In my twenties it was always very fraught with tension. When I was young and the stakes weren't very high—that is, I didn't care if the relationship ended up in a permanent union—there was always the thrill of the new, the excitement of going out: whom you would meet, who would hit on you, whom you would go home with, and who might hang around for a while. Every time you walked out the door there was a chance something could happen, and pretty often it did. Many times it ended in disaster.

Of course the sex was fun. I think that is what makes young dating so memorable—the sex is hot, even with people you don't like all that much. But if you did like someone, there were those endless next days and weeks of "Will he call? Should I call? Why didn't he call? Should I have slept with him? Should I not have slept with him? Should I have ordered fish when he said he was a vegetarian? Should I have ordered steak when he said he was from Omaha? If I'd worn the red dress, maybe he would have called. If I hadn't talked about my last boyfriend so much, maybe he would have called. If I were a cat lover and not a dog lover, maybe he would have called." It was endless. It was torture. It eventually tainted the sex. It eventually tainted everything.

Once you hit thirty, if you aren't married, the stakes suddenly ascend to colossal heights. I was part of the generation that was clobbered by the 1986 *Newsweek* headline that decreed women forty and over had a better chance of getting blown up by a terrorist than of getting married. Twenty years later they retracted the statement, but my guess is that faulty piece of journalism was responsible for more hasty, doomed marriages than we can possibly imagine. There is no question I was one of the headline's casualties. By the time I hit thirty I was so terrified I would be one of those left behind and I was so sick of dating that I married the wrong man just to remove myself from the scene, stop the insanity, and guarantee that I wouldn't end up one of those desperate forty-year-olds without a mate and forced to wear a flak jacket.

By the time I was thirty-nine and my marriage was intolerable, I realized I needed to get out of it by forty in the hopes of making a new life with someone else and possibly having another baby. I work well under deadlines, so I set a very strict one for myself. I had to be out and single by my fortieth birthday, and I was, mostly; I was somehow not single, but I was with the person who is the love of my life. I knew that if I did what many of my unhappy friends were doing, which was to hang in there and stay in the marriage for my daughter's sake, by the time she was out of the house I would be fifty and my chances of making a second life for myself, while possible, would be more difficult, and conceiving another child would be out of the question. So I did the selfish thing and left my first husband. This caused my daughter to accuse me of "hogging all the happiness," but I knew that if I was miserable for a decade, I would only end up making her equally unhappy, which would be merely spreading around the misery.

I knew then that there was a big difference between dating at forty and dating at fifty, the main one being the pool you have to choose from. I often advise my younger friends that if they are unhappy, they should get out now. (I actually give my older friends the same advice.) Why waste a day of your life being unhappy? It's not healthy for children to grow up with unhappy parents in an unhappy household. It means that all they will experience in terms of the male/female relationship is one that doesn't work. And oftentimes they will end up mimicking this in their own lives.

Kids are tougher than you think. Yes, divorce leaves a scar; yes, it's awful to go through; yes, it would be better if you didn't have to make the choice and the family could stay together, content and complete. But if that is not an option and if it becomes a choice between two unappealing situations, I feel it's better to go with at least a peaceful and fight-free household. Kids know, they pick it all up; you may think you are fooling them, but you never are.

And another thing: they have their whole lives ahead of them, while you have to move now to ensure that you have as many chances as possible. It might sound selfish, but at fifty you have to start calculating how many years you have and how you want to spend them.

For many women the thought of being alone at fifty and beyond is more frightening than sticking it out in a dead-end situation. Despite that, many women end up alone at fifty anyway, sometimes of their choosing, many times not. Numerous changes happen in relationships at this time of life.

Some women get left. It sucks, it's unfair, but it happens every day: some fifty-five-year-old guy decides he wants a thirty-year-old and a new family—in part, I think, just to prove that he can (plus it makes him feel younger). He figures hot sex with some-one new is better than the comforts of a long-term relationship, a family, and half his net worth, even if he has to put up with potato sex from time to time. (I think the potato sex does a lot of damage.) And the fact we can't pop any more buns out of the oven at this point in our lives sometimes makes us the less desir-able of the species. But even if you're having great sex and things appear to be going fine, just like with health, in relationships, shit happens. As one of my friends describes it, "My husband left me for a twenty-five-year-old. One of the worst things was being a statistic, a cliché. BTW, life is a process of finding out that you're not that special, that whatever can happen to anybody else can happen to you. People would say to me, 'How horri-ble it must be for your husband to leave you for a twenty-five-year-old.' I would say, 'It would have been horrible if he had left me for a forty-one-year-old Jewish therapist. What he wanted, I didn't have.'" Truly, if each woman could really believe that her unique gifts will be appreciated by the one who's supposed to . . . I used to say, "If you like my type, you'll love me, 'cause I do me really well."

This is a very important way to look at yourself as you move forward from fifty on: "I do me really well." Doing you is all you can do. And more likely than not you will be exactly what some-one is looking for. But to make it happen takes time, effort, and sometimes patience.

Being left for someone younger is one of the big ceilings that can fall in on your head at fifty. As if things weren't crazy enough, now you wake up and find yourself alone. It's scary. It's lonely. And it puts you back out there on the movable meat rack known as dating.

If your man leaves you, often it has little or nothing to do with the way you look. I know great-looking women in their forties and early fifties who still get left; as a friend points out, "If a guy wants what a twenty-year-old has to offer, he no longer wants you. Chances are he's a major dick anyway, though for many years you pretended not to notice."

There are other reasons women end up alone. Some marriages wear out but they stayed together just for the kids; once the kids are gone, there is no reason to keep up the facade. Some women are widowed. And some incredible, brave women wake up and realize they are not where they want to be—what worked for them at thirty or forty is not working for them anymore, and they go out there and brave the sometimes harsh, occasionally cruel winter of being a single woman in their fifties.

I believe it is better to be peaceful alone than miserable with someone. I have been there, and being unhappy when you are actually in a relationship is lonelier and more depressing than simply being on your own and actively doing things to make yourself a new future. Sadly, many women don't understand this and stay in bad relationships because they figure it's better to be with someone than with no one. Unless it is financially impossible to break free, it really is better to get out, especially when you are in your early fifties. You can choose to be happy, and happy does not always mean a man and a relationship. (Though a man doesn't hurt, unless of course he drives you crazy.)

In your fifties there is still time to start over. Look, people build new lives in their seventies; I wouldn't want to have to do it, but it is done. You're not going to see your golden anniversary, but if you start from scratch in your fifties and if you both live out your life expectancy, you can be married for thirty years. It's not the majority of the stories out there, but nearly every week in the *New York Times* wedding pages you see a couple in their fifties. If it can happen to them, it can happen to you.

Is it going to be as easy as it was at thirty? Of course not, but you knew that going in. The biggest problem for women in their fifties is that men their age often want girls twenty years younger. (Yet another case where fifty is *so* not the new thirty.) You can be the hottest fifty-year-old in the state, yet as far as men are concerned, you are still not thirty. The biggest complaint I hear from women my age is that they feel "invisible." I completely relate to this: you get to a certain age and men just stop paying attention. It's as if you do not exist at all. You go from guys turning their heads to get a peek at you when you walk into the room to no one even bothering to look in your direction. They are oblivious to how potently their indifference affects you, but every non-glance is a reminder of the ones that came before and the fact that they are no longer coming your way.

(I miss the construction workers whistling and yelling, "Sit on my face, blondie." Despite the fact that I always studiously ignored them, I secretly enjoyed the validation. Now when I walk by construction sites all I hear is, "Hey, Nico, will you get me a sub and a soda?" Ah, the old days. What I would give for one "Sit on my face, blondie.")

However, one of the good things (at least I think it's a good thing) about being this age is that you have so little tolerance for bullshit. I just can't imagine putting up with the crap men dished up and I took when I was younger. Now, after one "I will call you tomorrow" that didn't happen, he would be deleted from my contacts before the day ended. And I would not obsess over it, I would not sit by the phone—well, I guess we don't have to sit by them anymore; we all have them with us. We carry our potential rejection wherever we go. Which I think makes it worse. I see this with my daughter—the lack of immediate communication or validation keeps young people in a constant state of agitation. At least in our day, you had to wait to get home to confront the big red zero blinking on the answering machine. You had hours to pretend or fantasize that he might have called. Or you could find a pay phone and call home to hear, "You have no messages." (If we go back to the previous decade, it was the answering service: "Nothing tonight." And that was more humiliating, as the person giving you the bad news knew you were waiting for something, anything, other than "Nothing tonight." And they always said it with a note of sympathy in their voices, sort of like when the shrink says, "Your session is over" and adds mentally, *I know you need more, but you are not going to get it*.)

I think many women, especially those who have been married and have children, wouldn't mind being in a relationship but are not willing to put up with as much as they once were in order to have it. This all falls under the heading of "I don't care what people think as much as I used to, and I don't have to fit a certain stereotype anymore. I can just be me now, and that is freeing." It is a big bonus of getting to this stage. We have proved a lot of what we want or need, and can lighten up on certain things, like having to have a man. If you have been married and have kids, you beat the bogus *Newsweek* scare, so the ticking clock is not an issue. You don't have to get married; you've proven someone wanted you, even if it didn't last.

But humans still crave companionship, connection, love, and, despite the diminishing hormones, sex. So most people who find themselves single for one reason or another eventually take a deep breath and go back out there for another round. I know how tough it is; my single friends all tell me—constantly. You never walk up to a fifty-plus single friend and casually ask, "So, seeing anyone?" It borders on the cruel. If she's seeing someone, you will hear about it.

While talking to women about the brave new world of dating, my curiosity got the best of me. Most of the single people I know, from thirty to sixty, have dabbled in the land of Internet dating. And this ranges from the really hot properties to people who might not have as much luck on the open market. We all have a few friends who have found love through their mouse and keyboard. (My sister-in-law met her husband of twelve years on a music lovers' website back before everyone was signing up for insta-mates.) So with my husband's blessing, and in the name of research, I decided I would take the twenty-first-century approach to finding the perfect man: I would live like a single fifty-year-old online and see what was out there. I would sign myself up for one of the eight hundred online dating services.

If I did find myself single, I would surely end up taking a stab at this, so why not give it a shot? I'm not the only one: roughly forty million American singles use online dating and social networking sites to meet new people. Three million Americans have found long-term relationships or marriage through a dating site. One of those sites, eHarmony, has estimated that it was responsible for 43,051 marriages last year, which averages out to 118 marriages a day. This means that a total of 236 eHarmony users got married every day. Seventeen percent of online daters (nearly three million Americans) have turned online dates into a long-term relationship or marriage; one in eight people married last year in the United States met online.

To put it in perspective, in 2008, the last year for which stats are available, 34,017 people died in car crashes. So your chances of getting married from a hookup on eHarmony outweigh your chances of dying in a car crash by 26.5% percent. Those are pretty decent odds.

I went with Match.com. For some reason it felt like I was reaching for Kleenex instead of the generic brand. Despite its success rate, eHarmony sounds like an a cappella group, and it costs a lot more than Match.com, plus it feels like you have to fill out more forms to sign up with them than you do to join the CIA. J-Date is very popular, but though I was born a Jew I don't practice, and since I was already swimming in a pool of lies, I didn't want to travel further into the deep end.

I'm not easy to please under the best of circumstances, so I went in with many reservations. But as I posted on my profile page, "I am curious and open for new adventures." The first turn-off was when I couldn't use my regular name as my user name; it was way taken. They offered me three others: Tracey67548, which sounded like one of those early AOL addresses or a prison number; Hotmamita, which gave off too much of a Latin vibe; and PaltryTracey. This last was an odd choice, considering the dictionary defines *paltry* as "worthless, trivial, miserable, insignificant, and irrelevant"—the only adjective they missed was "invisible." I wondered if this had something to do with my age. Did the fact that I was fifty and online looking for love automatically render me miserable? The truth is, 30 percent of the eighty million boomers are single; they can't all be paltry. And what does that say about my sense of self? Would you want to date a girl—pardon me, woman—who described herself as irrelevant? I decided to stack the deck against myself. PaltryTracey was online and looking for love.

The photo I tried to upload was far from Paltry (I figured that if my name was going to be paltry, my image should be anything but); it was one of my press shots, and apparently for Match.com it either had too much cleavage or was too airbrushed. I decided it was the airbrushing, as it's not that revealing and I use it in Indian publications and they are very sensitive to exhibitionism of any sort. My guess is they put the photos through some sort of bullshit detector to see how Photoshopped or air-brushed they might be, sort of like scanning kids' IDs in clubs to see if they're fake. But I could be wrong, as most people on the site look pretty damn Photoshopped to me. So it might have been the cleavage, but a lot of women are up there in bikinis, many who shouldn't be, so perhaps it was just an upload error in the system. Whatever it was, it took a day to get it sorted out.

The second major turnoff for me was the fact that in the world of online dating, fifty is considered a senior. While I'm not thirty, I am no senior. (And not to brag, but one night not so long ago, one of my daughter's nineteen-year-old guy friends said I was "hot." Granted, it was dark, but that still made my decade and was far from paltry.) But Match.com classified me as a paltry senior and put a black bar over my boobs while my photo status was pending. This was not a good start.

At least I got the picture sorted out soon enough (Match.com ended up accepting a photo that was a little sexier but had not been retouched, so I think it was just a matter of the pixels). And I will say this: the men out there move fast. In the world of online dating, once you sign up, you are the new girl in town—fifty-eight guys looked at me within the first eighteen hours. Four were interested, and one winked; I got a message from someone saying, "Nice smile." The winker was the best of the lot, but he lived in Michigan. He was a widower; I was to learn over the four weeks I spent online dating that widowers seem to be the most sincere, and the ones I felt the guiltiest about lying to.

I didn't want to get into many conversations with people, though I did respond directly to one widower in Tennessee who had a little boy. He emailed me and I felt bad; he looked like a nice man. I told him he lived too far away but that I wished the best for him and his son. But then I got back an odd email saying we could work out the geographical issues if we loved each other. I then noticed he had listed his age preference as between forty-two and sixty-nine, which seemed odd considering that he had lost a wife to death; why would he want a sixty-nine-year-old to help raise his seven-year-old son?

I did not feel invisible any longer; I felt like an imposter, and that I was part of something slightly surreal, but invisible I was not. That may be one of the more appealing aspects of online dating: from the moment you sign up, you are back in some form of the game. You can go out every night without leaving your desk or getting out of your PJs. People you never would have the opportunity to meet come flying into your home via the computer. Over a period of four days, 329 guys looked at my profile. I was being checked out! But out of those 329, only twenty-five thought enough of me to follow through. Only twenty-five found me worthy of a second glance. I'm one of those people who if I get three compliments and one criticism I focus on the criticism, so true to form, as opposed to looking more closely at the men who supposedly were interested in me, I had to go prowling through the more than three hundred who had rejected me. Whom was I not good enough for? Many were way older, some even in their late sixties; one was a mortician, and one of them actually referred to his favorite vacation spot as "Canncoon." If nothing else, I could have helped him with his spelling, but I guess guys don't roam these sites looking for tutors.

Once I decided that the ones who'd rejected me were not really my type either, I could focus on who was interested and what my real chances might be if I'd been taking this seriously. Of the twenty-five who responded, seventeen winked and eight sent messages. The winkers are a promiscuous group who don't seem to pay a whole lot of attention to where you live or what you might be looking for. But then I guess the very nature of winking is benign promiscuity, especially if winking doesn't even require moving your eyelid, only clicking on the send button. My winkers were mostly around forty-eight years old, and out of seventeen winks, eleven were from out of state, some as far away as California and Texas. I'm not sure what these guys get out of it; maybe it leads to phone sex. Who knows? But why is some forty-one-year-old from San Diego winking at me? Where can that relationship lead? Maybe they're charity winks; they figure a fifty-one-year-old who calls herself paltry will be grateful for anything, even a fifty-five-year-old who can't spell *Cancun*. (Not that I made it onto his list, as I had to keep reminding myself.)

The winks actually mean nothing; it's the messages that count. Those are the guys who want some contact. All the ones who actually wrote to me were in their late fifties to mid-sixties, and despite the fact that I specified I lived in Manhattan, only one Manhattanite responded. It was mostly guys from New Jersey and Long Island, plus the nice widower from

Tennessee. The message here seemed to be that for the most part New York City men do not want to date paltry seniors, though I think I knew that going in. But the fact that they were in the age range I was looking for, and that most were fifty-year-old men who wanted fifty-year-old women, was very encouraging, even if there were only 8 out of 329 and one of them couldn't spell. (*No—he didn't want me; I must accept this. He saw my photo, read my bio—and did not want to take me on his next trip to Canncoon. Get over him, Tracey. Wait, I'm happily married. How did I fall down this rabbit hole?*)

Some of these messages come your way because the guys viewing you connect your mutual points of interest as a way to see how you measure up to each other. Which actually makes sense, except the only things I had in common with any of them was my exercise regime, the fact that I like to eat out, and that I don't smoke. I'm sorry, these are not interests; they are survival techniques. It's like saying we're meant for each other because we both like to wear coats in the winter. You have to eat to live, smoking kills you, and exercise helps keep you strong and healthy. (If any of you had any qualms about jumping into that daily exercise habit, this should give you incentive. It was the only thing all my prospective mates responded to, even the forty-year-old winkers.)

The one thing I emphasized in my profile was my sense of humor; I told them I was "wickedly funny." Which, call me egomaniacal albeit paltry, I think is true: I am funny. Nobody responded to this. Well, one did, but only to question it. He wasn't sure I was funny, he said, but I sure was beautiful. "You should see me airbrushed," I wanted to respond. I'm not beautiful, and even if I were, I will be funny a lot longer. And then, who is this guy? I mean, here he was questioning my sense of humor when he wouldn't even post his job status online. All of a sudden I felt like Lorne Michaels was auditioning me for *Saturday Night Live*. So despite the fact they all claim to "love fine wine, enjoy walks on the beach at sunset, and want someone to share laughs with," clearly they needed to be the ones providing the laughs. Or perhaps they figured that if I was too wickedly funny I would make jokes at their expense, which of course is exactly what I'm doing right now, so perhaps most of them were smarter than their pictures suggested.

(Which leads to another problem I found: many of the guys looked like they'd just gotten off work-furlough programs, at least the ones who came through in the beginning. My guess is these are the ones who are glued to the computer 24/7, and the second someone new arrives, they "check her out." Wink wink.)

I must say, the men who were interested in women of my age were rather particular. (Many had these peculiar age cutoff points, like fifty-two; what's that about? Say you're going to turn fifty-three in a month; does that take you out of the running?) I cut a very wide swath for my guys. I said I would take anyone from forty-five to sixty-five. That's the thing about us girls: we are willing to go way up in the age department. And I suppose if I really were to take this seriously, I would go up to seventy. But at fifty, I don't want a seventy-year-old, unless he's Robert Redford.

Another issue I discovered in the online dating world is that there is too much information off the bat. One sixty-two-year-old, also from New Jersey, really liked Aerosmith. In the absolute there is nothing wrong with liking Aerosmith. But the idea of this bald former pilot doing air guitar to Aerosmith was such a turn-off I immediately clicked "not interested." Now, of course we know I'm not interested anyway, but let's just say I was, and we got to know each other and took walks on the beach and shared coffee and conversation. (They all seem to be into coffee and conversation; it's the Starbuckification of our culture, the cheapest date. Better to spend $3.50 and forty minutes than $50 and two hours. I only wish there had been Starbucks when I was dating.) But anyway, imagine we got to that point where we had more in common than eating, protecting our lungs, and push-ups, and then, after I was thoroughly smitten, I found out he liked Aerosmith. That I could put up with. I would be so wowed by his other traits, like hang gliding and whipping up gourmet treats for two, that by the fateful night he happened to start jamming to "Livin' on the Edge," there would be enough feelings and shared moments between us that I would be able to get beyond it.

But if I know it going in . . . well, I ain't going in.

It's like the guy from Weehawken who likes Andrew Lloyd Webber. I have such an aversion to Andrew Lloyd Webber I cannot begin to describe it. I could never date anyone who liked Andrew Lloyd Webber, no matter how "charming and considerate" he might be, or how vibrant the sunsets we watched while sipping "really chilled wine." Even if the sex was mind-blowing, even if he bought me a house in India, every time I looked at him I would hear, "Nothing can harm you. My words will warm and calm you, I'm here. Let me be your freedom," and I would end up running all the way from Weehawken back to Manhattan. But if we met and I was captivated by that magic element that is often the glue of really powerful relationships, I could learn to wear those sound-blocking earphones or make him use them while he was listening to *Cats*. But having access to someone's minutiae can often turn you off before the real person has a chance to turn you on. Perhaps part of the problem is being a "senior"—you just don't have the patience for all the shit you find intolerable, even if it's something as banal as the score from *Phantom of the Opera*.

One guy looked okay on paper, despite the fact I was six years older than his cutoff point. He said he did a lot of charity work (great!) but loved billiards. *Next.* Another professed a love of tight blue jeans and posted a photo of the bottom half of a guy's body wrapped in skintight jeans. *Gross.* (Clearly a double standard here—if I was scrutinized for a smidgeon of cleavage and some airbrushing, why does he get to show his junk?)

Maybe it's why most guys want girls in their thirties: they're more tolerant—as long as you give them those babies. That is the one thing, aside from an aversion to smoking, that all the guys who checked me out shared: an equally powerful aversion to having kids. Most already have them.

Is it that the further on you get and the more entrenched you are in your ways, the more difficult it is to compromise? Absolutely. For example, I could never date anyone who has a TV in his living room. That is right up there with Andrew Lloyd Webber, but many guys photograph themselves in their living rooms in front of a big TV. For many women it's probably a turn-on: "Wow, he's got a eighty-six-incher." Me, I go on to the next one.

Mr. Canncoon did not do this; he was photographed on the beach in—where else?—Canncoon.

You can tell which guys and girls (I checked out my competition) have been doing this for a while; they know exactly the right mélange of photos to exhibit. The first is usually a close-up, in which many wear sunglasses. I totally would have done this if I'd known I could. What fifty-plus person doesn't look better in sunglasses? So you look at this tan guy, who may be standing nonchalantly near a sports car (we don't actually know if it's his or if he works on the lot), and he's in some cool shades; he looks good. Then you press forward and he's in his house, minus the shades, near the TV. Doesn't look like the same person. There are then several shots of him doing various sports to prove how young, in shape, and adventurous he is, and if he has them, a few travel shots are obligatory. Most of those are on the beach, if they aren't the goofy ones in which he's standing straight with the Leaning Tower of Pisa as the backdrop. These say, "Come fly with me and there will be some great travel in our future." And then the ones who want to come off as kind usually have a shot with a dog, never a cat. Real men don't have cats.

I noticed many of the women had shots in hot tubs or pools, where they looked tanned and also sportive. They had a shot in a black cocktail dress, resulting in one guy asking if there was a single dress that all the women shared. Some of the men had photos in evening clothes, showing that they could clean up well, but more had photos in shorts and flip-flops. For me, shorts and flip-flops are up there with Andrew Lloyd Webber and a TV in the living room.

If I were really serious about this I would do an album that showed the many facets of my personality: me reading a book, me typing, me reading a magazine, me shopping, me arguing with my teenager. I guess I could drum up a few shots in a Margaritaville somewhere in the Caribbean with a giant drink and a plate of nachos in front of me to show what a fun-loving girl I really am. That might have gotten me a wink from Mr. Canncoon.

I did send out one wink myself. Despite all the jokes, there were a few guys that, were I on the market, I could see myself going out with. There was one in particular who made me think: *Yeah, if I were single, this would be someone I could date.* First off, he looked a lot like my husband; also, he lived on Long Island, which isn't my first choice, but it's more doable than Dallas. He seemed interesting, and there was irony in his self-description— real irony, not something he picked up from a magazine; he's the one who made the remark about the shared cocktail dress. I liked that. It seemed to say that he got it. And there was a picture of him in jeans drinking red wine, which is what I look like every night around six-thirty. He also read many of the same books my husband reads. If I were doing this for real, those would have all been signposts that told me I had more in common with this man than a mutual love of gnocchi al pesto.

But there were also a lot of pictures of him skiing. Skiing was clearly a big part of this guy's life, that and boating, so he was way more sportive than I am. My bookshelf is more sportive than I am.

He did not respond for two days. I figured this was because of my lack of snow bunny shots. (I realize now that there is a cat-and-mouse aspect, just like in real-life dating; you don't respond to a phone call immediately so that you don't look too needy, and the same goes for cyberwinks.) But he did get back to me; he thanked me for the wink and called me a little tease. I didn't want to touch that one, but I figured that since I'd started this, I might as well see how far I could take it without making a total fool of myself or hurting anyone's feelings. So I sent back a snarky response; he nailed me on that, then said we should talk on the phone. Guess he didn't need a skier. Guess he liked snarky. Maybe this is the way this whole thing works. It was getting sort of fun.

I asked my husband if I could phone him. He said. "For the book? Sure." (His first wife moved away for a year to write a book; what was one phone call to Long Island?) I didn't want him to have my number, so I called him, but as my nine-year-old pointed out, he must have caller ID—so he could Google me and find out who I was and that I wasn't for real. But it was too late by then; I had called him. We chatted for a few minutes and all I did was lie. I hated myself; it was the day before Thanksgiving, and he was baking pies—or so he said. He seemed to be honest. He sounded nice; if I weren't happily married, I might really like him. I imagined him making pies alone in his kitchen; maybe he was only looking to get laid, but I don't think so. (He wasn't getting the real me, so it was hard to tell.) He seemed to want to talk again. I told him I was going away. He said we would talk when I got back. We didn't. I was so elusive that any connection would have required me to make the next move, and frankly, I was so noncommittal and boring on the phone that if I were him, I wouldn't have wanted to see me.

The next day a New York lawyer who listed Harvard and Yale on his resume sent me a long, sweet, funny email. He lived in New York City; he wanted someone my age. I figured that if he wasn't a serial killer, some girl would probably find him to be a nice guy. And then a really cute guy not only winked but sent me a message. His user name was FStopFitzgerald; he was a photographer, he had a sense of humor, his photos were cute, and he was fifty-three. If I'd been looking for real, I'd have been at Starbucks in a heartbeat meeting him for coffee, if he suggested such a thing. I actually came clean with him—he seemed too cute to lie to, and he was what I had said I wanted: someone artsy, someone funny, someone living in New York. He was only two years older than I was, he looked younger than his age, and he seemed to be in a higher income bracket than me. In the totally today way he and I ended up as immediate Facebook friends, with a friend in common.

There are guys out there, ladies. There are.

The one thing about Internet dating is that it takes time; it's hanging in there until the right one comes along. Friends who have found mates tell me, "We exchanged emails for six months before we met." Like in life (well, I guess it *is* life), if you're looking for quality, you need patience. FStop said it's easy to find "dates but not mates." People are leery. If you are this age and online looking for love, no doubt rejection has visited you many times in one form or another. You get to the point where you are so sick of the games. If the men on the other end have had bad marriages, which most have, then they are usually emotionally exhausted and burned out. If they have come through ugly divorces, they are terrified of going through that again. Some have lost their mates to death; these are the ones who seem most willing to put themselves out there a little more. They say that no matter what age you are, if you have had a good relationship, you are more ready and willing to jump back in; while there is sadness, there is far less fear.

And for some guys it's just a big frat party. I actually knew a guy who ended up marrying someone he met online. But he played around online a long time before he met her and settled down. He told me that if you are a guy with a few bucks cruising the dating sites, it's like a candy store: you can meet and nail three girls a day, coffee at Starbucks and then bingo. They don't even know your phone number or where to find you to yell at you if you don't contact them afterward. Several guys actually used the term "candy store" to describe it. Everyone I spoke to said it was the easiest way to get laid.

A friend of mine is forty-eight, attractive, and single. She has two kids and has done both old-fashioned fix-ups and Internet dating. She was very articulate about it and feels that at this age (and despite the sex.com aspect I've just described) people want some peace. They just want to know that whoever they are hooking up with is going to be there for them in whatever lies ahead. She feels that this very romanticized notion of relationships that many people write about on the Web is not projection but reality. They really do just want to walk on the beach and snuggle in front of the TV; they want to wake up and share the paper and a good cup of joe. At fifty, despite the shots of them on the ski slopes and the all-terrain vehicles, the guys looking for women their age are looking for security and safe place to come home to, she says. I tend to agree with her; it's not always easy to find just the right person. And then, of course, not all the guys are like this; there are the seventy-year-olds who demand someone no older than thirty-five, and the fifty-five-year-olds who just want to get some easy pussy. Like the rest of life, it's complex and virtually impossible to categorize easily. Great relationships involve a certain magic that is hard to find. Good ones sometimes are the result of patience and a dose of settling for perhaps a little less than you were hoping for. But that happens in the real world as well as in cyberspace, and I think the prospect of settling for a good if not great relationship is often more acceptable the older you get.

While the majority of the men I came across were not for me, they seemed like a decent lot of fellows. And despite the fact that I have an aversion to putting a TV in the living room, many women don't. What I took away from visiting the land of cyberlove is that there are many more real people out there than I could have imagined. And just as you can't find a mate in four days when you're twenty-five, you certainly can't do it at fifty.

There are opportunities out there to be had. (Had I been looking for real, FStop and the guy from Long Island might have been possibilities.) But as is true in other parts of our lives as well, at fifty we cannot have the same expectations we once had. There will certainly be compromises. The guy for you might not be in the next town; he might be in Michigan, and those guys in Michigan may know something New Yorkers don't—the world does not begin and end here. For love, many people will travel very far, and at this stage in life, the trip may be well worth it.

I got off the Internet scene quickly, as my very presence was leading people on. And not that I was breaking any hearts, but at this stage of life every rejection, even an imagined one, causes a tiny ping. Just as at first I felt rejected by Mr. Ski Slope and Mr. Canncoon, I'm sure the guys who emailed me but who didn't get an answer felt that for whatever reason they too weren't good enough. At this juncture we need to surround ourselves with life-affirming friends and lovers—the people who want the "me" I do so well.

The thing I find the most poignant is just how many truly

lonely people there are. And like many things, dating at this age is harder for women; we all know men are allowed to age in ways that we are not. And when you are brave and put yourself out there, how difficult and humbling it can be.

But no matter what their age, almost all people want to love and want to be loved, even if they have a thing for Aerosmith or a TV in their living room or aren't great spellers. People want to find love so badly that they actually believe that a shared appreciation for classic rock, rock shrimp, and rock climbing somehow makes them soul mates. And, as I learned from my brief foray into the world of cyber meet-and-greet, blessedly, it sometimes does.

10

Ready or Not, Here Death Comes

I don't want to achieve immortality through my work ...
I want to achieve it through not dying.
—WOODY ALLEN

When I was thirty I had lost one significant person in my life to death: my grandfather, who died at the age of seventy-seven. Looking back, it's quite amazing he lived that long considering the abuse from food and lack of exercise his body lived under and the fact that he had his first heart attack in his mid-sixties. Though I mourned him and miss him to this day, we're geared up for the fact that grandparents are going to die at some point ahead of us. It's in life's rule book. I lost my father's mother to lung cancer when she was also in her mid-seventies, which, quite frankly, is another accomplishment considering that she smoked a pack or more a day for sixty years. If Grandma Dot was death by Crisco, Nana Dottie was death by Kents. Had either my grandfather or my chain-smoking grandmother taken better care of themselves, I'm sure they would have made it to their eighties, or at least I would like to think so.

But at thirty I could count on one hand the people I knew who were my own age and who had died from something besides AIDS. (By taking AIDS out of the discussion, I am in no way discounting or minimizing it. During a five-year period in the eighties I did lose quite a large number of friends and countless acquaintances. It was a scary time and almost seems like an aberration today. Today the face of AIDS has changed, and if you have access to the new drugs it has become a manageable disease; take one look at Magic Johnson. I can honestly say that in the two decades since then, I have not known anyone who died of AIDS.)

The people my age who died by the time I was thirty went in the way most young people do—accidents. A boy I knew in school was killed in a car crash, as was a close family friend's son who was a shade older than me. The very first person my age to die was ten; she drowned in a lake at summer camp. It is beyond bad luck and for the most part incomprehensible to a young child. It didn't keep me out of the water, though I have never spent much time in lakes. I remember not so much fearing that it would happen to me—at that age death is far too distant a concept—but thinking: *What will her parents do without her, and how must her sister feel?* We weren't so close that I missed her, though I never could look at her house again without thinking of her at the bottom of a lake.

The death I remember most vividly was that of a friend's older sister, who went down in a small plane flying to Colorado. I was twelve then, and eight years later another girl exactly my age who had gone to our school died in the crash of a private plane. This led to a lifelong refusal on my part to step anywhere near small planes. I have to this day, despite countless chances, never been inside one, much less flown in one. If I don't get a boarding pass, I don't get on. And I have no fear of flying in the absolute. After 9/11 I was one of the first people I knew to get in a plane and fly across the country. But because of those two deaths my fear that small planes were deathmobiles far outweighed the über-deluxe status the rest of the world seemed to bestow upon them. If I stayed out of private planes, I could avoid one of life's arbitrary causes of death.

We spend much of our lives keeping death and disease at arm's length. And why wouldn't we? It's another reason we seem so set on this delusion that we're twenty years younger than we actually are. In fact, when we hear about someone's misfortune or illness, often the first thing we do is somehow try to justify it or ascribe it a reason unique to the other person's circumstances that is totally separate from our own. We do this not out of cruelty but out of fear. You hear that someone has lung cancer and your first remark—or, let me put this on my shoulders, my first question—is "Did he smoke?" If the answer is yes, we can take the cruel hand of fate out of the picture. And we usually respond with, "Thank God I never smoked" or "I quit years ago."

Thus we feel we will somehow be spared and the victim is, sad to say, the architect of his or her own misfortune. The same is true with breast cancer; whenever I hear someone has breast cancer, my first response is "Did her mother have it? Does cancer run in their family?" It is a totally self-indulgent and somewhat ignorant perspective to take, as only 10 percent of cancers are genetic. We just try so hard to pretend that tragedy somehow can't end up on our doorstep.

At fifty the death of people your own age or close to it is the one thing that really takes you by the neck and gives you a good hard shake. The sudden heart attack, the undiagnosed aneurysm, the cancer that metastasized before it was caught and thus carries a quickly terminal diagnosis, or the cancer that lingers and whittles away people still in the prime of their life—these are realities we are forced to face when we hit the half-century mark. It is one of the worst aspects, if not the worst, aspect of living 18,250 days. I'm not saying that people and families don't face death and disease and other cruel acts of fate before fifty, because they do. But the bodies don't start piling up with the same speed and frequency—at least not in my experience and that of my friends.

When you are younger, death is truly random. They were drunk, or made a bad choice and got in the car with someone who was. Fate, luck, or genetics were just not playing on their team. But after fifty, death happens and happens and happens and happens some more. I was not prepared for this.

In a period of six months, between my husband and myself we lost nine people. They include two of our closest friends, one or two people who would fall into the category of "close enough to have over for dinner, but not close enough to tell your problems to," people we knew through work or the gym, friends of friends, parents of friends. That is one and a quarter people a month.

The first to go was one of my writing students with whom I had become friendly. Friendly enough that she had been over to the house, friendly enough that we had met for coffee, friendly enough we had one of those hanging lunch dates that was always going to happen but never quite did. She died within six months of being diagnosed with uterine cancer at the age of thirty-six. She was fourteen years away from the big five-oh. With this, there is no way you can play the "it couldn't really happen to me" game. The truth is, uterine cancer is much more common in women over fifty. Cancer is by nature a degenerative disease. The older we get, the more we degenerate; ipso facto, cancer is closer to our door with every passing year. The weird thing is that her Facebook page is still up. Like several of my friends who have died, she still lives on in cyberspace, which gives death a whole new visage. Everyone gets the celebrity treatment as they live on in their vibrancy, but it's truly confusing to those left behind. I think someone needs to write an etiquette book about what to do about dead people in cyberspace.

The summer of 2009 started with Michael Jackson dying. Now, I didn't know Michael Jackson, and I wasn't his biggest fan—I had never really seen the magic in his moonwalk, and I had some issues with his personal life—but in a way I always thought of him as that peppy little kid, the one of the Jackson

5 who had a shitload of talent. In many people's minds he was forever that little Afro-wearing boy singing "ABC." Even after he morphed into his own version of Diana Ross and we probably all suspected he was doing something that was not too good for him in the psychopharmaceutical department and maybe his obsession with little boys wasn't kosher, he was still young, and though he might have been an odd choice he was an icon for our generation, the boy we grew up with who never grew up. He was one of us and then he was gone. His death we can also justify as out of our orbit, since most of us don't use anesthesia to get to sleep, but still, he was fifty, he embodied our generation, and now he's gone.

Our personal death list continued when my husband's spinning teacher dropped dead in the gym. A healthy, fit man—well, hell, the guy taught four spinning classes a day; try to do that and not be fit. He was fifty-one, two years younger than Glenn at the time. They had been discussing nutrition and the many attributes of spinning the week before. One of the most positive aspects of spinning is that it strengthens your heart, thus extending the quality and length of your life. So the spinning teacher who is two years younger than you are is not supposed to drop dead on the gym floor while everyone drips sweat as they try to revive him. This flies in the face of everything every boomer fitness addict believes in—that if I do this, I will beat the odds.

Glenn, who is not shaken easily and is nowhere near my league in the hypochondria department, came home slightly traumatized. He was not close to the teacher; he barely knew him. But Glenn, like many people including me, believes that if you exercise, you are doing the best thing you can to keep yourself healthy, strong, and alive.

This was a "fifty is truly on the way to old age" moment, and while you can spin to your heart's content, it still may give out on you. Glenn hadn't even changed out of his sweaty gym clothes when he called and made an appointment with the cardiologist. That week he had every heart test that has been invented. He came home and described them in great detail. I heard about every vessel, artery, and chamber. He had an MRI, as well as a CAT scan where they can see every artery and how badly they might be blocked—a diagnostic tool that I think can really alert you if danger lies ahead. They upped his Lipitor and his blood pressure medication. He swore off meat and salt (that lasted about two days). And then he did something people love to do—he played God with someone else's life: "If the spinning teacher had only taken these tests, he would be alive today."

Since justifying the spinning teacher's death was nearly impossible, he did the only thing he could; by being overly medically responsible, he removed himself from the scary fate of the spinning teacher. The guy may have been in great shape, but he didn't take the right tests.

"And Tim Russert," he bellowed.

"Tim Russert what?" I said.

"If Tim Russert had had these tests, my tests, the tests I just took, he would be alive today." Clearly Tim Russert's death had been weighing on him, too.

"How do you know he didn't?"

"Everyone at the hospital said he hadn't. He had a stress test—that only tells you if you have stamina. My tests tell you . . ." Suddenly they were *his* tests, Glenn's tests.

"But maybe Tim Russert did have your tests."

"There is no way. If he had had my tests, he would have lived."

He was starting to sound like he had not only taken the tests but invented them as well. How my very level-headed husband had talked himself into this place I have no idea, but it proves there is no limit to the mind games we will play in an attempt to distance ourselves from a death that is too close for comfort. And Tim Russert's death was easier to rationalize than the spinning teacher's. He was overweight, he was under a lot of stress, and every time you saw him he was eating a hot dog. (Well, I saw him only once in an airport—but he was scarfing down a hot dog. He loved baseball, and people did talk about his love of hot dogs.) But not everyone who eats hot dogs drops dead in their fifties. Tim Russert was a brilliant, beloved man, but a fitness fanatic he was not. I don't pretend to know what Russert's medical history was or what tests he did or did not take. But somehow Glenn made himself feel like he would not keel over randomly on the StairMaster, as he had done the responsible thing—he took all the right tests.

But even having the heart test of all heart tests clearly didn't ease his discomfort over the death of the spinning teacher. He stopped taking spinning. And then one day several weeks later he came back from the gym triumphant. "Preexisting condition! The spinning teacher had a congenital heart defect." There it was: the real-life "it can't happen to me" mind easer.

"Really? Who told you?"

"Some of the guys at the gym."

This had become the topic of the over-fifty set at his gym over the summer. I guess guys who had not taken "Glenn's tests" had to come up with some reason they too would be spared death on the gym floor. So a congenital heart defect became the group's pacifier. Apparently several of them were so obsessed by this and spent so much time talking about it that the owner of the gym asked them to take it outside, as it was making the people who hadn't witnessed it nervous.

People drop dead of heart attacks when they are young as well—every now and then a young basketball player goes—but it's so infrequent and such a freak act of nature that you can write it off to bad luck, or sometimes to "He did a lot of coke off the court." But at fifty, death starts happening more and more: to your friends, your favorite newscaster, or even the guy who is supposedly teaching you how to avoid it. And it scares the shit out of you no matter how many ways you try to distance yourself from it.

The other major death of the summer, the one that brought me literally to my knees, was my best friend, and I mean my *best* and oldest friend. We had been friends for forty-eight years. We met on the playground in nursery school. We went through grammar school together, spent holidays together, ended up in the same field, and wrote films together. He was the person I was closest to besides my husband and children, though he had known me longer and better in some ways. And he died suddenly of a pulmonary embolism on August 4, three months before his fifty-second birthday.

When he died, a part of me went with him.

We looked alike; we knew more about each other than anyone else knew about us.

Our collective joint experiences spanned our entire lives. As an older friend said the other day when she was describing losing her best friend, "She was my memory. She'd been in my life so long she remembered the details I had forgotten and I did the same for her." There are few people in anyone's life one can say that about.

In true twenty-first-century fashion, I learned about his death on Twitter. One of the most earth-shattering experiences of my life, reduced to a 140-character tweet.

Is this what it means to be fifty—your best friend dies of a pulmonary embolism?

Younger people die of pulmonary embolisms. In fact, my GP, Dr. Bernard Kruger, says, "It is an ageless condition brought on by a multitude of different causes." So this could have happened when we were in our thirties, but it didn't—it happened in our early fifties. It got lumped into my new life of being fifty. I totally connected the dots.

His death was so close I couldn't separate myself from it,

nor did I want to. It was inside of me, a part of me; it lived and breathed with me. His death defined me for months after I found out about it. His ceasing to exist was something I would have to adjust to over time, yet I will never get over it. It was hard knock number one in the "beginning to lose your close friends" phase of life. And considering who he was and the role he played in my life, it was a horrible place to start.

That same week we lost another close friend, though he had been dying for the past year and he had reached the age of seventy-two—not old by today's standards, but old enough that you don't really have to come up with all the reasons it won't happen to you. His absence will be a big loss in our lives; he was always first on the list for celebratory parties and family occasions. He will be deeply missed.

At the end of that week Glenn and I looked at each other and said, "Welcome to our future." The summer pretty much continued in that vein, though the other deaths were not as close or alarming. But it seemed like every week there was another, including people we didn't know but felt we did. John Hughes, the boomers' cinematic yearbook editor, died of a heart attack while taking a walk in Central Park. Farrah Fawcett and Patrick Swayze died of cancer, both of them boomers in the Bermuda Triangle of health.

The last week in September 2009, as I was on my way out the door to attend a friend's father's funeral—he left behind two sons in their fifties—the phone rang. It was my stepmother telling me she was following the ambulance, she didn't need me to fly out yet, but my father was in critical condition and might not make it. So I sat through the funeral wondering if I might be attending my own father's next. Blessedly, he got to the hospital in time and pulled through, though had it been an hour later, they say, he would not have made it.

This is the other hard-core reality of how fifty is so not like thirty. Half of fifty-year-olds have lost at least one parent. If they haven't lost one, they are taking care of one, worrying about one, or very likely dealing with the health issues of one or both.

Both my husband and I still have both our parents. My father comes from hearty stock, and if family history is an indicator, he could be around until ninety, which will put me in the group who get to sixty with a parent. My mother, I'm not so sure. Since we are no longer close, her death, whenever it comes, will bring with it a whole load of other issues to deal with. That's a place I don't want to even venture near yet. It's hard enough to lose parents, but when you lose them and still have holes that have been left unfilled, issues unaddressed or not sorted through, and lingering bad feelings from either party, it has to be an awful way to face the future. I know many people for whom this is a reality. I fear it will be a reality for me and one I do not look forward to. They say you should never go to bed angry with those you love. How about taking off for the hereafter with unresolved conflict being the controlling dynamic? I think that could actually be a legitimate definition of hell for both sides, the departed and the one left behind.

Yes, yes, yes, we know death is part of life. We know all you can count on is death and taxes. And you always hear, "Ashes to ashes, dust to dust . . . the wheel of life . . . nothing lasts forever . . . he lived a great life while he was here . . . she's gone to a better place . . . we were lucky to have him for the time we did . . . let's be grateful for what was and not focus on what isn't." These are all useless statements that I think trivialize a deeply traumatizing experience. Of course in the absolute they are true. But even if you believe in the afterlife, which I for one do, it still doesn't make living on planet Earth easier without those you loved.

How about "We will all be together someday"? Okay. I can go there because it's more comforting in the moment than the other option, which is that I will never see the people I love again.

But what if you're at the Olive Garden and I'm waiting for you at Sizzler? What if I come back a goat and you come back a pop star? We don't know. We pretend we do because we want to control all aspects of our existence, whether in life or in death. But all we really know in these moments of loss is that there's no one there to forward that funny email to. That your parent may never get to hold her grandchild or see him graduate or get married. That you can't ask for advice—or be given it even if you don't want it. A person you found to be a pain in the ass while he was alive leaves a crater in your heart once he is gone. The person who could fill in your memory gaps is no longer there to do so for you. The lifelong best friend whose phone number is on speed dial in your brain is no longer at the other end of the line, or waiting in your usual spot for coffee, or there to share holidays, celebrations, new books, old movies, the latest restaurant or just what it was like to be you two together for all these years; that person is now gone from this life forever.

When you get to your fifties, you can and do make new friends—good friends, from what I've experienced. But at fifty and after there is no way you can accumulate the same collective mileage you have with the people who have come before. No one else will ever know me the way my friend Blake did because no one else will ever have known me as long or gone with me through so many stages of my growth. The frames of reference, the shorthand that makes up long-term friendships, cannot be duplicated with new people. And that is one of the huge losses you start experiencing at fifty that you never did at thirty.

Whether you liked them or not, whether they lived up to your expectations or not, you get only one set of parents, and once they are gone, that is it. And in that loss, along with all the grief, the complications, the memories and memorabilia, is the knowledge that in your family tree, most likely you will be the next branch to snap off.

Since I haven't experienced this yet, some very generous friends shared their feelings with me. One friend writes, "The hard days are easier since my parents died. After standing there and watching each of them die, no bad day can compare to the sadness and pain of that slow gurgling out to their last breaths. What sucks? The really good days. When one of my kids has an amazing achievement in school, gets a brilliant review in the press for her performance, finally starts to hit the ball in Little League, I want to pick up the phone and tell the only ones who would be completely, unselfishly ecstatic for my kid. There's no Mom to call. There's no Dad whose email address I can type into the 'to' line as I cut and paste the review of my child's show from the paper. It's the loneliest, saddest part of my life. Just when I'm the happiest for my child's moment of glory, that emptiness comes rushing back. It's just not the same with 99 percent of my friends. My friend C. is the only one. *She* feels the same way. We have a deal—we share this kind of stuff with each other because we both miss our mothers so much and in the same way."

Another friend shares this note she wrote to colleagues when her mother died: "It is with great sadness, relief, guilt, shock, guilt, relief, and hurt that I tell you that my mother died suddenly on Monday morning. I tried to see her as my Zen master teaching me patience and forgiveness, and as you particularly know, I was not even close. Who's going to push my buttons? Once again, I'm learning the difference between theory and experience. Experience hurts more." She signed it, "J. the orphan."

A friend from Texas: "I feel that death is never a welcome visitor for those left behind. I remember both my parents' deaths as if they were yesterday. My father died on Memorial Day in 1988. I was forty-two. And my mother died April 16, 2007. When my father died he already had one leg amputated and they were about to take his other one. I was at the hospital in Miami and the image of the empty stripped bed is a sight I will never forget. The foam egg crate filling they put on his bed was all that was left. I fell to my knees in pain. By that time he was my best friend and I was lost. To this day I miss his 'Hi, pal' on the other end of the phone. It's not many times that I hear the word 'pal' used in my everyday life, but every time I do, my thoughts go to him. My father passed his sense of humor on to my sister and me. When I think I have said or written something funny, I secretly thank him. It has become my coping mechanism. As we all know, death usually buries the bad times and seems to add a little extra seasoning to the good times. That's fine for me."

A friend compares loss at thirty and loss at fifty much better than anyone else I know. "Before I had turned fifty I had lost two members of my immediate family: a sister who died violently well before her time and my mother, who succumbed to a long and painful illness. I was not a stranger to mourning, so when my father entered the last stages of cancer in his eighties, at the end of a rich life worthy of biographies, I felt prepared, even accepting of the inevitable. My loss is now only a few months young, but I'm well enough into the process to know that Kübler-Ross did not have me in mind when she formed her now famous five stages. I was stunned with unexpected depression to not only lose my father but be revisited with the loss of my mother and sister. This triumvirate of depression could only have happened had I not lived over half a century. It was as if I was dealt these feelings because I had dodged the sadness of the previous deaths and was now old enough to take on my father's passing and handle the other two as well. I had been given an exquisite triptych of grief to deal with on my own, given time. Bring it on, bring it on."

I don't think I understand much about death, and I'm not sure anybody really does; it's part of what makes it so terrifying. We don't know when it will hit, whom it will hit, and at what speed it will be traveling. We have a preconceived notion of the order in which people should die: grandparents, parents, then us. But the pain of losing our parents can take us by surprise—the guilt, the missed opportunities, the fact that our heart skips a beat every time the phone rings and we know it won't be them—and it takes getting used to. And then there is the other harsh reality of losing your parents: you are supposed to be next. It doesn't mean you will be, but the house is betting on it.

Plus there is the uncertainty: as we begin to slowly lose our friends, the next twenty to thirty-five years become a macabre game of musical chairs. Who will get to sit down for another round and who will be eliminated? There is nothing to do about that except pay attention and when the music stops grab the nearest chair.

And then of course, as clichéd as it is, at twenty or fifty or eighty we have to try as hard as we can to think of each day as a gift, love each member of our family even if they drive us crazy at times, and cherish our friends the best we can—and never take our own life or anyone else's for granted.

11

The Flashlight at the End of the Tunnel

The best is yet to come.
—Cy Coleman and Caroline Leigh
(written when he was thirty-five
and she was thirty-seven)

When I was thirty I was convinced I would never make it to forty, much less fifty. There was no real reason for this—just a lifelong negative mind-set, or perhaps a childhood spent watching too many episodes of *Medical Center*. When I was ten, I never thought I would make it to fifteen. I like to set myself up for disappointment and then be pleasantly surprised. The other positive about gearing yourself up for disaster is that you're inclined to be prepared. If you're always waiting for the other shoe to drop, you tend to keep an extra pair in your bag.

Now that I'm past fifty, I'm hoping to make it to my late eighties or nineties—provided I am in good health. I have stopped with the "I won't make it to the next big birthday" phase. When you're younger it comes off as idiosyncratic; in your fifties it seems like a stupid game to play and totally counterproductive. Though none of us likes the idea of getting older, the alternative is a lousy one, so we must trundle forward with as much dignity and enthusiasm as we can muster.

Unlike Mr. Coleman and Ms. Leigh, I do not feel the best is yet to come. And it's important to note that they wrote the song in their mid-thirties. Caroline Leigh, who wrote the lyrics, died at the age of fifty-seven, so when she was fifty the best was not yet to come for her. Cy Coleman went on to be seventy-five and certainly had many great years ahead of him. He was nominated for his last Tony Award at the age of sixty-eight.

So you have two different scenarios for the way things can turn out between fifty and eighty. One gets sucked into the Bermuda Triangle of health; the other goes on to a vital, productive last lap, working into his seventies.

While I don't think the *best* is yet to come, I do think many great things lie ahead in the next several decades. But for the most part the events are toned down from what we have gotten used to in the last thirty years, and many of the big moments will belong to others and we will find some of our happiness in being a part of theirs. Our children will start and hopefully flourish in their careers as ours wind down. They will marry or find their mates, while many of us will end up living alone or in relationships that do not have the vigor they once did, though many people lucky enough to grow old together find a new type of closeness in their later years. We can look forward to grandchildren who will visit, but not to children of our own living at home.

All of these things, I am told, are glorious moments and bring great joy. People tell me that being a grandparent trumps parenting any day. They all say the same thing: "They go home." But if you are one of those parents who enjoyed having your kids at home, I'm not so sure an endless parade of hellos and good-byes is as satisfying as the daily activity of an on-going dynamic family life. My grandmother always hated to see me go, and my mother did too. The good-byes always got to them though they tried not to show it. These are still not the moments we were rehearsing in the Fisher-Price house so many years ago. Did you make a pretend pot roast only to have your baby run out the door because she had other plans? Do you ever remember arguing about who got to be the mother of the bride and who got to be the grandmother? Or who got to spend Thanksgiving with you and who went to her spouse's family?

"You get to be the stepmother-in-law whose name is not on the wedding invitation and who refuses to show up at the rehearsal dinner."

"Okay, I'm the mommy, you're the daddy. You're fifty-six, and you just got fired because your company is downsizing. Our oldest kid doesn't speak to us because her fiancé says we are controlling, and our youngest is in rehab. My mother is suffering from dementia, but we can't afford the good nursing home and rehab at the same time, especially with you not working. My

best friend just got diagnosed with ovarian cancer and her husband just came out of the closet, leaving her alone and humili-
ated at a time when she really needs someone."

I was a very paranoid kid, yet even I didn't play house that
way. We always played "the best is yet to come," and why
wouldn't we? Who wants to even think about the other version?

At this point, if we look down the tunnel of life and don't
admit that things are going to change and at times become very
rocky and more difficult, we are deluding ourselves. There is no
blazing light down at the end anymore. There is a light, but it is
a flashlight; it beams strongly at moments and flickers at others.
Sometimes it's as bright as a klieg light, but then it starts to dim;
it can and frequently does beam brightly again, but over time it
gets softer and softer until eventually the batteries wear out and
the light no longer works.

At fifty, the stats are in our favor—the majority of us have
many decades to go, but they will not all be like the ones that
have come before. If you have gotten this far and had a fairly
normal life, chances are the worst is yet to come in many ways.

The worst is yet to come in terms of your health; there is no
way around that. We are decaying, we will die, and the longer
you live the greater your chances for a protracted death as opposed to an instantaneous one.

Our parents will die, if they have not so already. If they are
alive, we will have to see them through illnesses and say good-
bye. If we have strong relationships with them and few regrets,
we will miss them, but we had good years and will soldier on

with lovely memories and hopefully much of their wisdom to hold on to and pass onto our own kids. Not everybody gets this, so if you do, be grateful. If you don't—better luck next life.

If we are fortunate enough to live long lives, we will end up burying many friends. My friend Buck calls it our "punishment." That's a little dark for me. Though I'm not sure how you really deal with it; we just do.

All of us will face some health crisis or another. Twelve percent of us will have a mammogram day that does not turn out well. Two hundred and sixty-seven thousand women will die each year from heart attacks, which kill six times as many women as breast cancer; eighty-three thousand of them are under sixty-five, and thirty-five thousand are under fifty-five. That heart scan doesn't seem so crazy when you start crunching the numbers.

Sixty percent of married women will lose their spouse and thus be left to live alone for a period of time, often until the ends of their lives. And, sadly, it's frequently a period when we don't have as much to keep us busy and distracted, so we have to make things happen for ourselves.

Our jobs will peter out. Even Cy Coleman, who had a forty-year career, ended up not working as much if at all in his final years; and he was as big a star in his field as they come. Life winds down, and we have to be prepared to wind down with it.

What I'm talking about is coming to peace and aging gracefully. It's not giving up, it's not throwing in the towel and looking like shit and feeling worse; it's understanding that you had your time to excel and now life is going to start drifting in another direction and you'd better drift with it.

People in India have a theory that I think makes sense (even if they don't practice it as much as they used to). They feel the first third of your life is for the accumulation of knowledge; the second part is for bringing new lives into the world and caring for the family; and the last third should be spent in contemplation of God, in preparation for spending so much time with him in the future. It is a comforting theory; spiritual beliefs can get us through tough times, and they often become more important the older we get. I'm not prescribing organized religion to anyone—that is each individual's choice—but the older you get and the more loss you face, the more it can sometimes help to think you have a benevolent friend somewhere out there.

I find that women who are further down the road seem to have an easier time. They are more at peace, and if they are in good health, they don't seem as demanding that life be what it once was. They are far enough away from youth that they accept it's gone. But those of us who are around fifty right now are still too close to our younger years to truly separate from them. Some of us still have remnants of it in our lives in the form of careers not yet changed, younger children or children who are just now leaving, parents still living. Many of us are still living parts of our youth. If we have taken care of ourselves and done some tweaking, we might not look like we're fifty. We blessedly still have the majority of our friends and our health is still good. The real bada-bing-bada-boom hasn't happened, but we have glimpses of it all the time and we don't like it, so we pretend it isn't so.

Here we find ourselves again between a rock and a hot place. We aren't old and we aren't young; we are in a kind of in-between state, passing through the transit lounge of life. We're not ready to let go of what we were in the past, and often we are not yet asked to. But we are also not comfortable accepting the future that sits in wait around the corner. Can you blame us?

I think the loss of a certain power is what terrifies so many. The fact is, you can't just bounce back from a job loss or setback the way you once could. In many areas we don't have the options we did in our early years and we don't get the do-overs like we did in our thirties and early forties. Our health is more precarious. And we watch friends attacked by so many enemy diseases that it raises our own health alert to orange. We are in probably the biggest state of flux we have ever been in, with perhaps the least appealing parts of life ahead. No wonder someone came up with the brilliant notion of shaving off twenty years.

But by shaving off those years we just set ourselves up for failure. If you look at yourself at fifty and tell yourself you are thirty, you will be confused and depressed more often than focused and joyful. Playing thirty at fifty is a game we cannot win. We have no choice but to stare fifty down and meet it on its terms.

There is a Buddhist parable I am very fond of and try to remember when I need it. Three monks have to get to the top of a hill. A very large, angry, man-eating dog lives on the hill.

The first monk heads up the hill only to be met by the snarling beast. Out of fear the monk runs from the dog, but the dog chases him, catches him, and eats him. The second monk takes the same journey; he too runs from the dog, and the dog chases him down and eats him. Then the third monk marches up the hill. He meets the dog, and as usual, the dog snarls and bares his teeth. But this monk runs straight at the dog, so terrifying the animal that it heads off in the other direction. The moral, of course, is that if we run toward our fears, they lose their power and ultimately disappear. I think when it comes to aging this is a helpful parable. It says that we shouldn't give in, let ourselves look like a wreck, move to Florida, and count down the hours until the early bird special begins. It encourages us to accept the limitations we are faced with, work with them, get past the ones we can, not let them get to us, and never give up.

There is a Dick Wolf quote about the film business I always liked, and I called on it many times when the thought of giving up was much more appealing than hanging in there. He said, "The person who makes it in show business is the person who stays in show business." I think we can apply this directly to life: the person who makes it in life is the person who stays in life.

The great thing about being in our fifties is we have this last splash of time. We are not senior citizens, despite the cards from AARP or the fact we can start collecting our pensions or move into senior housing developments. We have a good decade to sprint to sixty and make some serious changes in our lives that will set us up for that last phase. And I think most people at this stage realize that what we do now really counts. This is why you find people at fifty abandoning relationships and jobs that aren't fulfilling. People make big life changes at this stage because they realize it really is now or never. There are things you can still do at fifty that you will not be able to pull off again. I think fifty lights a fire under the brave and the committed.

Since no one really gives us a blueprint or a practice run for this, we have to figure it out as we go along. But we can, and many of us do. We have to run toward aging, and while it won't run away, we can make it something we don't hate, and make ourselves the people we want to be in this new stage of life.

People always ask kids what they want to be when they grow up. Nobody asks, "What do you want to be as you grow older?" I suppose for many the answer would be "alive." But being alive and really living are not the same thing.

By owning that I am aging and consequently asking myself, "Who do I want to be when I'm older?" I have been able to compile a list of the things that I think make for a vibrant, satisfying last lap. I have also picked out several women I admire—ranging in age from the late sixties to the early eighties—and have made mental notes on what they do, how they do it, and the choices they have made along the way. It has given me a blueprint as to how to proceed from here.

One of the main components involved in getting from here to there contentedly is having as few regrets as possible. I find that the people who by their fifties or early sixties have accomplished what they set out to do in their twenties and thirties are the ones who can look back and say "I did it" with a sense of pride and fulfillment. They are the ones who can let go more easily and move on to something new with more enthusiasm and fewer if any regrets. Having lived the life you wanted when you were young almost always leads to a happier older age. Living with regrets inevitably leads to "If only I had . . . ," and "If only I had . . ." is far too often the express train to anger.

When my film career ended and Hollywood and I mutually decided we were sick of each other, I could honestly take stock and say I'd accomplished 90 percent of what I set out to do. I missed the giant blockbuster, my $100 million comedy, but otherwise I hit every one of my targets. When I look back now, I do so with no regrets. And when I see the state of the film world, I am grateful I was in it back then and am not trying to start out today. That makes going forward so much easier. But if you can't do that and there are things you really want to do, you need to put them on your bucket list now and get going. The whole *Eat, Pray, Love* phenomenon resonates because there are so many things people want to do but they never take the time or muster the courage to go after them. No one's going to do those things for you. It's up to you to sit down, figure it out, and get a move on.

It may be too late to become a heart surgeon, but it's not too late to help start a small free medical center for the poor in your community. It may be too late to be a ballerina, but not to open your own bakery. Many people have back-burnered certain passions while they were making money and raising their families; now is the time to do those things. If you are in a miserable relationship at fifty, why would you hold on to it for another thirty years? Let it go, move on, choose happiness while you can. Having regrets at fifty is okay; you still have time to undo many of them. Regrets at sixty and seventy lead to bitterness and a "who gives a damn now?" attitude. Which is why so many older people are such bummers to be around. They are looking back and going "Shit, I didn't do half of what I wanted. What the fuck, it's three o'clock—let's go get dinner."

Look at life as though it were a trip to Hawaii: pick the sights you want to see and the things you want to experience, and then do them before your plane leaves for home. You may not fit them all in, but try for the important ones. There is nothing worse than heading to the airport from a vacation and saying, "Damn it, we missed the waterfall. That was the one thing I really wanted to see." Like many destinations (and depending on your beliefs), you won't be coming back here, so get it in now—fifty still leaves you time to do it.

The women I admire and am modeling myself after have all managed to have both big careers and families. While each one is alone now (two are widowed and one is divorced), they all live active social lives—they go out, they entertain—and all of them surround themselves with people younger than they are. They do not label themselves as "older women." They are women who happen to be chronologically older, but they are not hanging out playing mahjong and comparing generic medications to the brand-name ones.

Each had a career, and each still works. One of them was as powerful in the business world as a woman can be. She ran a huge division of a multibillion-dollar corporation. She was a star by forty, so she was one of those who got to hang in there until sixty-five, but eventually she was edged out, and instead of going home and living off her savings and complaining, she is now starting up a new, very forward-thinking company. She is surrounded by young people who know more about certain things than she does, but she is showing them how to run a big business. She can't wait to get up every day and head out there to see what she can do. Will she ever run a multibillion-dollar company again? Who knows? She might be past that point in her career, though never say never. However, she will run a company, and it will grow and she will grow with it. Many of the problems and challenges she faced each day in her earlier career will be the ones she faces now on a smaller scale, but the real bonus is that she is not running someone else's company, she is running her own, and she won't have to step down until she decides to. She is in her mid-sixties and holds her future in her own hands.

She just became a grandmother and could not be more thrilled. In fact, she is turning one of her children's old rooms into a nursery. She is a force of nature and derives satisfaction from her own accomplishments rather than living vicariously through her children, and because of this they want to spend time with her even though they are now starting families of their own. I look at her and say, "Yeah, that's the way to be in your late sixties."

Now, not everyone can start a company the size of the one

she has, but this is a case where size doesn't matter. It's about new frontiers, enthusiasm, and not going gently and quietly into older age because the corporate world thinks you should. She is a corporate person, and she understands that world; now she is making her own world, one that will work for her. She is a major role model for me.

Another woman I greatly admire is in her late sixties or early seventies—it's hard to tell, and she doesn't. She has worked most of her life in academia and still does. While she doesn't carry the course load she once did, she still teaches, she is a prolific writer who continues to publish, and she is deeply involved in politics and feminist causes. The woman never stops. Between her work, the committees she chairs, the books she writes, and her love of travel, she is like a whirling dervish. She is alone in the sense that she does not live with a man, but she is out all the time, entertains, and loves music, art, and theater. She is always taking in new information and formulating new ideas. You never walk away from her feeling like, "Oh, poor thing, there she is by herself"; you walk away amazed that she manages to accomplish so much in the course of the day, and you can't wait to see her again to find out what new things she is doing.

She too is close to her family and just became a grandmother. When you stand back and look at her you say to yourself, "Okay, I can be that. That looks okay. It doesn't look like the young women sitting in the park pushing their three-year-old on the swings, but it looks like a good life, a complete life, a life being lived to the fullest."

Then I have my friend Joan, who at eighty-two looks sixty, acts forty, and lives like she is thirty. She has the generosity of spirit to tell people up front how old she is—she is proud of it, and she should be. She knows that for those of us who are not there yet but have certain fears, her mere presence and life force have the power to alleviate them, so she declares her age and gives us all hope and something to shoot for. She has been through her share of hard knocks and lost many people she loved. She is a widow and lives alone, but once again, you don't feel sorry for her. She still writes books and works for *Allure*—yes, at the age of eighty-two, even in these times when all magazines are letting people go, they kept her on. Her stories are invaluable to them.

She also didn't even begin her real career, the career she has today, until she was in her forties. She has turned herself into a star during her last long lap, and it's ongoing. So the "star at forty" thing can be proven wrong if you have enough talent, moxie, and drive to set the bar high and keep clearing it.

She takes impeccable care of herself; all these women do. She is always busy and trying out new things. She is starting up a new website, and she is presently making her first documentary film. She too bounds out of bed each day full of plans and activities and deadlines. It would be easy for her to just sit back and be lazy, give up, and complain; it would be easy for any of these women to do that. They have all faced the same challenges and disappointments and losses that others their age have. They've been fired, been widowed, lost friends, had health issues; one even lost a child. They've been through it all. But each one chooses to be active; they look forward, they keep finding new ways to grow, and consequently they make their lives work on their terms.

As we all know, no one can stop the clock, not through words, slogans, or wishful thinking. But by taking good care of ourselves, keeping active, always learning new things, giving back to others, utilizing the advancements in cosmetic enhancement if you desire, and remaining realistic yet optimistic, we can live full lives for decades. We may not be thirty when we're fifty, or forty when we're sixty, but we can be the best version of where we are at any given time.

We are at the place where we must acknowledge that the best may not be yet to come, but we must make the best of what *is*. And only by owning it and not bemoaning it will we be able to do this. There is a great difference between bemoaning and owning: saying we are getting younger is denying, and thus rejecting, where and who we actually are, whereas accepting where we are and doing something to make it better gives us back the control we feel has been taken from us.

The truth is, it's not your grandmother's fifty, and it's certainly not thirty, but it's *your* fifty. And though certain things will eventually come to a halt, the quality of your life doesn't have to.

We have that power, but we have to commit to it; we have to set targets and hit them. We have to keep moving forward. Run toward the dog. Not only is it great exercise, it's not nearly as scary as you think.

Acknowledgments

First there was William Goldman who told me if I didn't start writing books and stop devoting all my time and energy to screenplays, I would lose my mind. Plus, he was the only one honest enough to tell me that as a fifty-year-old comedy writer in Hollywood, I needed a plan B. He was oh so right.

Then there was super agent, the elegant Ed Victor who was convinced I could pull off writing a book long before I was. He in turn convinced the team at HarperCollins that I was worth taking a chance on.

Jonathan Burnham came along on his white horse and played Lancelot at exactly the right moment, and he handed me over to splendid editor Jennifer Barth who showed me one could edit, prune, and pluck a writer's work and make it better without destroying the garden she had created. After twenty years in Hollywood this is something I didn't think was possible.

Eventually I landed in the lap of social media maven Debbie Stier who took my hand and led me through the labyrinth of cyberspace and turned me on to the tools of online self-promotion.

There is Carl Lennertz who taught me about booksellers and how to handle myself amongst them.

Sandi Mendolson who kindly spent months fielding calls about my documentary before she was required to do anything and then kicked butt to get my name and the book out there.

Richard Arlook, my manager who has run the marathon of my career by my side. And no matter how many holes I fell into, he has always been there to help pull me out and set me back on course. The last few years he has seen little monetary compensation, but he has never for a second lost his enthusiasm, devotion, or belief in me—I am eternally grateful.

This book would not be possible without the contributions of so many. First, I have to thank all the doctors who not only put up with me but freely gave their years of experience and knowledge and the many who read over chapters in order that I did not screw up all the facts.

Dr. Edward Liu who I have loved from the moment I first saw him and who has been with me through so much for so long.

The plucky Dr. Robin Phillips who made my menopause go away. Dr. Jon Turk who by giving me new eyes helped me see the world in a new way. Doctors Bernard Kruger, Ariya Neilson, Ellen Gendler, Gary Solomon, JT, and BE; I promise I will never use your email addresses for anything medical.

Mary Caraccioli who came to my rescue when I had no sound financial advice to give.

Lynnda Blitzer, Tracy Wientraub, and Susan Stoltz Davis for sharing and Mr. Stoltz for being the perfect mom next door.

Kelly Langberg for leading me to Jon Turk and being such a good friend.

The other Tracey Jackson, who gets much of my email, even things I send myself. When I mistakenly sent her a copy of the book, she read it and suggested I needed a chapter of finance: How right she was.

The generous friends who relived the pain of their parents' deaths, and shared their memories with me so I could share them with my readers: Lizanne Rosenstein, Griffin Dunne, Judy Harris, and Larry Enzer.

And the women who have shown me by example how to live fully, productively, actively, and not be fearful as the years go by: Domna Stanton, Jane Friedman, and MamaFace Joan Kron.

Though he's sadly no longer with us to accept it, my deepest love and gratitude for the forty-eight years of friendship, advice, writing tips, and laughs—Blake Snyder. I promised you I wouldn't stop writing.

And finally there is nothing I can do without acknowledging my grandparents Dorothy and Phil Jacobson who were always there and in many ways always are.

About the Author

A screenwriter for seventeen years, Tracey Jackson has written and sold films to all the major studios; her most recent writing credits include *Confessions of a Shopaholic* and *Lucky Ducks*, a feature-length documentary that she also produced and directed. Jackson blogs on her own website and on the *Huffington Post*. *Between a Rock and a Hot Place* is her first book. She lives in New York City with her husband, Glenn Horowitz, and two daughters.